CORRUPTED SCIENCE

JOHN GRANT

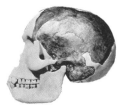

Vinlanda Insula
a Byarno repa
et leipho socijs
Mare Occeanum
Magnae
Insulae
Beati Brandani
Branzilae
dictae
Desiderate
insule
Mare Occeanum
Beate isole
fortunate

CORRUPTED SCIENCE

JOHN GRANT

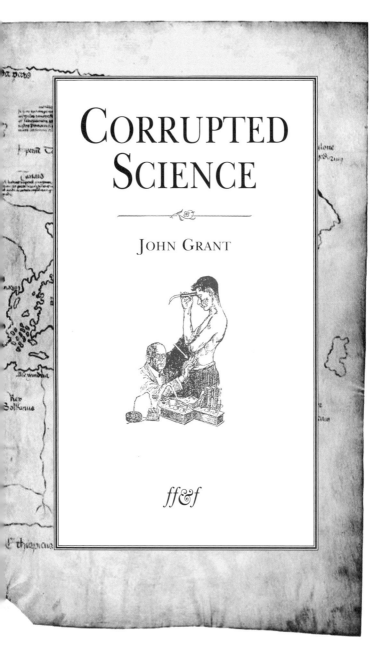

ff&f

DEDICATION

For Keith Barnett (1938–2006), the best big bro any little bro could
have hoped for. Oh, Keith, that you might have lived to see this
book complete . . . and to find my errors in it.

Acknowledgements

First of all, again The Spammers: Randy M. Dannenfelser, Bob Eggleton, Gregory
Frost, Neil Greenberg, Jael, Stuart Jaffe, Karl Kofoed, Todd Lockwood, Aaron
McLellan, Lynn Perkins, Ray Ridenour, Tim Sullivan and Greg Uchrin; between
them they sent me more material than I could possibly hope to use. Ian Johnson likewise sent me items of note; thanks, ol' buddy. John Dahlgren graciously permitted
me a month's-long extension on another deadline so I could finish this. Cameron
Brown had faith in the project and commissioned the book, while Malcolm Couch
created, yet again, an excellent design; the level of my gratitude to both is embarrassing. Lynn Perkins, out of the goodness of her heart, gave me the astonishingly
helpful present of a subscription to *New Scientist*. Joyce Barnett gave me an extremely valuable volume from the collection of her late husband, my brother Keith; even
greater was the gift thereby of his involvement with this book. Pamela D. Scoville
monitored various newspapers and online sources for new data I might have missed,
but far more significantly continued living with the general intolerability of the husband who was writing this book: *Je t'adore, ma cherie*.

A few paragraphs of this book first appeared in different form in my *A Directory
of Discarded Ideas* (1981) and *Discarded Science* (2006).

CORRUPTED SCIENCE

Published by Facts, Figures & Fun, an imprint of
AAPPL Artists' and Photographers' Press Ltd.
Church Farm House, Wisley, Surrey GU23 6QL
info@ffnf.co.uk www.ffnf.co.uk info@aappl.com www.aappl.com

Sales and Distribution
UK and export: Turnaround Publisher Services Ltd. orders@turnaround-uk.com
USA and Canada: Sterling Publishing Inc. sales@sterlingpub.com
Australia & New Zealand: Peribo Pty. michael.coffey@peribo.com.au
South Africa: Trinity Books. trinity@iafrica.com

A catalogue record for this book is available from the British Library.

ISBN 13: 9781904332732

Cover Design: Stefan Nekuda office@nekuda.at
Content Design: Malcolm Couch mal.couch@blueyonder.co.uk

Printed in Malaysia by Imago Publishing info@imago.co.uk

CONTENTS

Introduction:
The Falsification of Science 7

There is, it appears, a conspiracy of scientists afoot. Their purpose is to break down religion, propagate immorality, and so reduce mankind to the level of the brutes. They are the sworn and sinister agents of Beelzebub, who yearns to conquer the world, and has his eye especially upon Tennessee.

H.L. Mencken reporting on the Scopes Trial, *Baltimore Evening Sun*, July 11, 1925

They think that differential equations are not reality. Hearing some colleagues speak, it's as though theoretical physics was just playing house with plastic building blocks. This absurd idea has gained currency, and now people seem to feel that theoretical physicists are little more than dreamers locked away in ivory towers. They think our games, our little houses, bear no relation to our everyday worries, their interests, their problems, or their welfare. But I'm going to tell you something, and I want you to take it as a ground rule for this course. From now on I will be filling this board with equations. . . . And when I'm done, I want you to do the following: look at those numbers, all those little numbers and Greek letters on the board, and repeat to yourselves, "*This* is reality," repeat it over and over . . .

José Carlos Somosa (trans Lisa Dillman), *Zig Zag*, 2007

THE FALSIFICATION
OF SCIENCE

———————⟨⊛⟩———————

Science is a set of rules that keep the scientists from lying
to each other.
  Kenneth S. Norris, cited in *False Prophets* (1988)
  by Alexander Kohn

IT SEEMS HARDWIRED into the human brain that people in
general believe what they're told by other people. One
can understand why this should be so: verbal communication was presumably fostered initially as a means of
exchanging vital information among tribal members ("the
food's over there") and to members of other tribes ("if you
try to take my food I'll kill you"). Speech would have little
survival value for the tribe – and indeed little purpose –
unless the information it contained were true. We can guess
that the invention of the lie followed some little while after
speech was in widespread use – and a devastating invention
it must have been. Even today we naturally tend to believe
what we're told – scepticism is an educated response, not an
instinctive one, as demonstrated by the ease with which
parents can fool small children with tall tales. Similarly,
most of us tell the truth almost all of the time.*

For the most part it's essential for the smooth functioning of society that this twinning of truth-telling and belief
continue. Consider the simple social interaction in which

---

* Hence the effectiveness of the joke applied to many politicians over
the years:
  Q: How can you tell when he's lying?
  A: His lips move.

you ask a passing stranger for directions. Society would soon crumble if strangers habitually gave false directions, or if tourists habitually disbelieved the directions given to them by strangers. Of course, sometimes strangers *do* quite deliberately give false directions, either because they don't know the answer and have an infantile dislike for displaying ignorance or in the misguided belief that it's funny. (Sometimes, likewise, travellers disbelieve the genuine directions they've been given because "they don't make sense".)

The deliberate giving of false directions might be regarded as a small-scale demonstration of the inherent flaw in our natural assumption of truth-telling/belief. If one party, almost always the teller, disobeys the tacit rules of the game, the other is exceedingly vulnerable. Hence the effectiveness of false propaganda, as exemplified today by broadcasters such as Rush Limbaugh (see page 302) and the pundits of Fox News: they can tell whatever fibs they like secure in the knowledge that a high percentage of the audience will believe what they've been told; further, since those members of the audience who perceive the lie will soon go elsewhere for their information, the percentage of remaining listeners who're credulous tends to rise. This particular dishonest gambit is not a new invention, of course: it can be traced back throughout most of recorded history. False tales of the disgusting licentiousness of Cleopatra (69BC–30BC) were circulated in Rome to shape the citizens' attitudes towards the Egyptian queen. Lies concerning the sexual appetites and extravagance of Marie Antoinette (1755–1793) contributed to the onset of the French Revolution. And we all know the disastrous consequences of the early-20th-century antisemitic propagandist forgery *Protocols of the Learned Elders of Zion*.*

The same vulnerability in our social structure is of course exploited alike by the hoaxer, the forger, the con artist, the trickster, the prankster, the dissembling politi-

---

* This was actually a reworking of *Dialogue in Hell Between Machiavelli and Montesquieu* (1865) by Maurice Joly (1829-1878). Joly's original was a perfectly honest satirical attack on Napoleon III. All the anonymous forgers of the *Protocols* did was change a few names and details.

cian, the religious fundamentalist (and indeed the self-styled prophet), the television evangelist, the propagandist, and the straightforward liar. Often the efforts of these assorted crooks are to harmless and/or humorous effect but sometimes, as per the *Protocols*, they have enormously damaging consequences; various political and media denials concerning imminent catastrophic climate change, for example, may spell the collapse of human civilization.

Most hoaxes and frauds are outwith the purview of this book, in which we're concerned only with the sciences, but that still leaves us plenty to play with. We'll start with scientists themselves – scientists who for one reason or another have felt driven to fake things. In the second half of the book we'll look at those who've corrupted science not from within but from the outside.

Left: Louis Pasteur *Right:* The skull of Piltdown Man
Backdrop: Marco Polo

CHAPTER 1

FRAUDULENT SCIENTISTS

———————— ⟨◉⟩ ————————

THE PROMINENT UK surgeon Paul Vickers remarked in a 1978 lecture: "What the public and we [doctors] are inclined to forget is that doctors are different. We establish standards of professional conduct. This is where we differ from the rag, tag and bobtail crew who like to think of themselves as professionals in the health field." High words, and arrogant ones, but coloured in hindsight by the fact that just a few years later, in 1983, Vickers, in association with his mistress, was charged with – and later convicted of – murdering his wife Margaret, using "professional conduct" to do so: he poisoned her with an anti-cancer drug so that her death was initially attributed to natural causes.

Arrogance and hypocrisy are recurrent themes, although far from the only ones, in considering fraudulent scientists: it is as if some get so engaged in their own world that the external one becomes a sort of secondary reality, one in which events have an almost fictional status and where consequences need not be considered.

But there are plenty of other motivations for fraudulence in science.

In 1993 the physicist John Hagelin (b1954) – three times US Presidential candidate for the Natural Law Party – gathered together in Washington DC some 4000 transcendental meditators with the aim of reducing the violent-crime rate in that city by 25%. Unfortunately, the police records showed that Washington's murder rate for that year actually rose. Or did it? In 1994 Hagelin announced he'd done a *proper* analysis of the figures and, sure enough, they showed a 25% decline.

On the face of it, this might seem like an instance of the fraudulent abuse of statistics, but was that really the case? It seems unlikely. Far more probable is that Hagelin, unable to believe the results of his experiment, quite unconsciously read into the statistics what he wanted to see there.

The borderlines between fraud, self-deception, gullible acceptance of the fake, and the ideological corruption of science can be very blurred. In theory none of them should happen; in practice, all too often, they all do, sometimes in combination. Where does one draw the line between, say, self-deception and ideological corruption? The latter may be deliberate and self-serving, but it can equally well be a product of the same desire to have reality obey one's wishes that seems to have driven Hagelin to derive from those crime statistics a different message from the one the authorities did.

All of the categories of scientific falsification are dangerous: people can die of them – either directly, as with fraudulent cancer cures, or indirectly, as when distortion of science about supposed racial differences reinforces irrational prejudices and hence leads to, or prolongs, violence between peoples. The Nazis had their scientific "proof" that the Jews were subhuman, so murdered them by the million. In 18th- and 19th-century North America the White Man had "proof" of the subhuman nature of Black and Red races, so enslaved one and waged a campaign of near-genocide on the other. The hideous pseudoscience of eugenics not only helped fuel the Nazis' murderous spree but also, before that spree shocked sense into people, looked well set to initiate a culling in North America of the insane, the socially inadequate and, of course, those of "lesser" race.

We live in an age when the falsification of science, in particular the ideological corruption of science, has reached a new level of importance – the very survival of our species may be threatened by it.

Not all falsifications seem of such significance – although one could make the case that false belief in itself does considerable damage by way of a sort of intellectual pollution that hampers all our other efforts at progress, or simply by generating sufficient irrelevant "noise" that genuine knowledge becomes obscured. What may start as a relatively harmless hoax or fraud can be compounded through human gullibility or self-deception until suddenly it assumes an importance far beyond anything the original perpetrator could have conceived.

On a small scale, this is what happened in the late 16th century in one of the oddest spats in the history of science, the infamous case of the Silesian Boy. This child was born on December 22 1585 and was discovered, when his teeth grew in, to be possessed of a golden molar. How could science explain this?

The best-regarded hypothesis of the time was produced by Professor Jacob Horstius (1537–1600) of Helmstädt University, who was also the first to publish a book on the subject: *De Aureo Dente Maxillari Pueri Silessii* (1595). Horstius's hypothesis managed to conflate astrology with a medical belief widely current at the time: the notion that, if a pregnant woman allowed herself to be captivated with desire for something she saw, the next time she touched herself on her body she would generate an appropriate birthmark on the corresponding bodily part of her as-yet-unborn infant. Horstius claimed the astrological conditions pertaining at the time of the Silesian Boy's birth (under the sign of Aries with the planet Saturn in conjunction) were so favourable that the boy's body began to produce not bone (as teeth were thought to be) but gold. The gold had manifested as a tooth because the boy's mother, while carrying him, must have coveted something golden she'd seen – coins perhaps – and not long afterwards stuck her finger in her mouth. Horstius went on to theorize, obscurely, that the tooth was a sign from Heaven that the Turks would be kicked out of Europe.

Another book, this time by Regensburg physician and alchemist Martin Ruland (1532–1602), emphatically seconded Horstius's hypothesis. Others felt driven to write books in rebuttal, among them Duncan Liddel (1561–1613), who pointed out an elementary flaw in the astrology of Horstius's proposal: the sign of Aries falls in spring, not in December. The Coburg chemist Andreas Libavius (1555–1616) produced a book about not the phenomenon or the theory but the controversy itself.

It finally occurred to someone that it might be a good

idea to examine the boy and his tooth. It was at this point that it was discovered the whole affair was a hoax: the parents had jammed gold leaf over the child's molar.

Not many years later Francis Bacon (1561–1626) set forth the Scientific Method – which advocated the collection of empirical evidence before advancing hypotheses – precisely in order to avoid such embarrassments. For centuries people accepted the principle . . . but carried on much as before, believing what they wanted to believe and finding the "proof" where they could. It's something we still do.

The idea that at least some blind people can distinguish different colours by touch dates back at least to the 18th century; there is mention of it in Boswell's *Life of Johnson* (1791). At that time it was believed minuscule differences in surface texture were responsible for the appearance of different colours; although these variations were too fine to be detected by most people, the non-visual senses of the blind were known to be often more astute than those of the unimpaired. In different form, the idea re-emerged in the 19th century, this time the underpinning being the notion that the different colours generated slightly different temperatures. One practitioner of the art in England was a teenager called Margaret M'Avoy (1800–1820), who was all the rage in 1816. A peculiarity of her uncanny ability which puzzled people was that it would function only in the light; in darkness her fingers were just as blind as anyone else's. Her adherents helpfully explained that this was because everything looks equally black in darkness.

That might have been the end of the story, but in the 1960s, when the fad for parapsychology was at a peak in the West, reports emerged from the Soviet Union of various women who could "see" with their fingertips or other extremities. Several of the practitioners were soon exposed as frauds, and the credulous reports in Western media – including notably *Life* magazine – tapered off rather abruptly when something very obvious was pointed out: stage magicians had been performing precisely similar feats for generations. Yet, where those reports were retracted at all, it was with considerably less prominence than the original features had been given, and so it is still widely believed among the general public that the women's abilities were found to be genuine.

The notion of "the emotional life of plants" can be traced to the polygraph expert Cleve Backster (b1924), the founder in New York, after a career with the CIA, of the still extant Cleve Backster School of Lie Detection. In 1968 Backster published a paper, in the *International Journal of Parapsychology*, "Evidence of a Primary Perception in Plant Life", claiming that, by in essence hooking up a polygraph to plants, he had been able to show plants possessed a primitive form of ESP: they displayed a reaction to the destruction nearby of living cells. Money poured in to help him establish the Backster Research Foundation, whose express purpose was to investigate further the ESP abilities of plants. The initial results from these researches were very positive, and it seemed Backster and his team had made a great breakthrough. However, other researchers couldn't replicate the experimental results, and during the 1970s Backster's work was very thoroughly debunked. This did not stop the publication of several best-selling books on the subject of plant psychology – notably *The Secret Life of Plants* (1989) by Peter Tompkins and Christopher Bird. It seems Backster's claims may have been the product of the fairly common phenomenon whereby perfectly rational researchers can unwittingly, and despite all self-imposed safeguards against bias, skew their results to favour their preconceptions. There has been no sign, though, of the publishers retracting the related books.

These two cases lead to another important tile in the falsification-of-science mosaic: the role of the print and broadcast media, which are almost always eager to trumpet sensational claims of the extraordinary and then, when the claims are shown to be bunkum, near-criminally negligent in acknowledging the fact. This is frequently compounded in the modern era by the perversion that has grown up of the old idea of journalistic balance: the new *faux*-balance seeks to find a midway point not between two reality-based viewpoints but between a reality-based viewpoint (right or wrong) and one that is demonstrably false. The result in the minds of the audience – and, who knows, perhaps in the minds of the journalists too – is a fallacious perception that facts are somehow subject to debate. Yes, of course, the *interpretation* of facts is very often open to discussion, but the facts themselves are not. The attitude that everyone's opinion is equally valid, no matter their level of ignorance or expertise – and certainly no matter what the reality actually is – is lethally dangerous in some areas of science, most particularly at the moment the science related to climate change.

BABBAGE'S REFLECTIONS

In 1830 Charles Babbage (1791–1871), best known today for his early work on the computer, published *Reflections on the Decline of Science in England, and On Some of Its Causes*. In this little book's Chapter V there's a subsection on the genesis of fraudulent science that's as valid now as it was then – which, indeed, is almost a textbook-in-miniature of how fraudulence, deliberate or unconscious, can arise *within* the sciences. In the following extract, the footnotes are Babbage's.

There are several species of impositions that have been practised in science, which are but little known, except to the initiated, and which it may perhaps be possible to render quite intelligible to ordinary understandings. These may be classed under the heads of hoaxing, forging, trimming, and cooking.

OF HOAXING. This, perhaps, will be better explained by an example. In the year 1788, M. Gioeni, a knight of Malta, published at Naples an account of a new family of Testacea, of which he described, with great minuteness, one species, the specific name of which has been taken from its habitat, and the generic he took from his own family, calling it Gioenia Sicula. It consisted of two rounded triangular valves, united by the body of the animal to a smaller valve in front. He gave figures of the animal, and of its parts; described its structure, its mode of advancing along the sand . . .

The editors of the *Encyclopedie Methodique*, have copied this description, and have given figures of the Gioenia Sicula. The fact, however, is, that no such animal exists, but that the knight of Malta, finding on the Sicilian shores the three internal bones of one of the species of Bulla, of which some are found on the south-western coast of England,* described and figured these bones most accurately, and drew the whole of the rest of the description from the stores of his own imagination.

Such frauds are far from justifiable; the only excuse which has been made for them is, when they have been practised on scientific academies which had reached the period of dotage. . . .

FORGING differs from hoaxing, inasmuch as in the latter the deceit is intended to last for a time, and then be discovered, to the ridicule of those who have credited it; whereas the forger is one who, wishing to acquire a reputation for science, records observations which he has never made. This is sometimes accomplished in astronomical observations by calculating the time and circumstances of the phenomenon from tables. The observations of the second comet of 1784, which was only seen by the Chevalier D'Angos, were long suspected to be a forgery, and were at length proved to be so by the calculations and reasonings of Encke. The pretended observations did not accord amongst each other in giving any possible orbit. . . .

Fortunately instances of the occurrence of forging are rare.

TRIMMING consists in clipping off little bits here and there from those observations which differ most in excess from the mean, and in sticking them on to those which are too small; a species of "equitable adjustment," as a radical would term it, which cannot be admitted in science.

This fraud is not perhaps so injurious (except to the character of the trimmer) as cooking, which the next paragraph will teach. The reason of this is, that the average given by the observations of the trimmer is the same, whether they are trimmed or untrimmed. His object is to gain a reputation for extreme accuracy in making observations; but from respect for truth, or from a prudent foresight, he does not distort the position of the fact he gets from nature, and it is usually difficult to detect him. He has more sense or less adventure than the Cook.

OF COOKING. This is an art of various forms, the object of which is to give to ordinary observations the appearance and character of those of the highest degree of accuracy.

One of its numerous processes is to make multitudes of observations, and out of these to select those only which agree, or very nearly agree. If a hundred observations are made, the cook must be very unlucky if he cannot pick out fifteen or twenty which will do for serving up.

* *Bulla lignaria.*

Another approved receipt, when the observations to be used will not come within the limit of accuracy, which it has been resolved they shall possess, is to calculate them by two different formulae. The difference in the constants employed in those formulae has sometimes a most happy effect in promoting unanimity amongst discordant measures. If still greater accuracy is required, three or more formulae can be used. . . .

It sometimes happens that the constant quantities in formulae given by the highest authorities, although they differ amongst themselves, yet they will not suit the materials. This is precisely the point in which the skill of the artist is shown; and an accomplished cook will carry himself triumphantly through it, provided happily some mean value of such constants will fit his observations. He will discuss the relative merits of formulae he has just knowledge enough to use; and, with admirable candour assigning their proper share of applause to Bessel, to Gauss, and to Laplace, he will take *that* mean value of the constant used by three such philosophers, which will make his own observations accord to a miracle. . . .

. . . it may happen that, in the progress of human knowledge, more correct formulae may be discovered, and constants may be determined with far greater precision. Or it may be found that some physical circumstance influences the results (although unsuspected at the time), the measure of which circumstance may perhaps be recovered from other contemporary registers of facts.* Or if the selection of observations has been made with the view of its agreeing precisely with the latest determination, there is some little danger that the average of the whole may differ from that of the chosen ones, owing to some law of nature, dependent on the interval between the two sets, which law some future philosopher may discover, and thus the very best observations may have been thrown aside.

In all these, and in numerous other cases, it would most probably happen that the cook would procure a temporary reputation for unrivalled accuracy at the expense of his permanent fame. It might also have the effect of rendering even all his crude observations of no value; for that part of the scientific world whose opinion is of most weight, is generally so unreasonable, as to neglect altogether the observations of those in whom they have, on any occasion, discovered traces of the artist. In fact, the character of an observer, as of a woman, if doubted is destroyed. . . .

---

* Imagine, by way of example, the state of the barometer or thermometer.

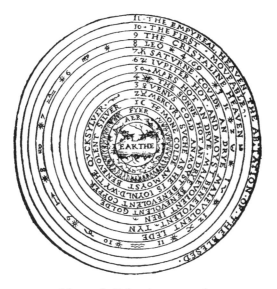

Schema of a Ptolemaic cosmography

CHEATING

Even very distinguished scientists are not immune to the temptations of fraud. For some 1500 years the Western world regarded the geocentric cosmology of Ptolemy (c90–168) as the last word on the subject, and he was greatly admired for the way in which he had confirmed his theories by experiment. Even long after the Copernican Revolution outmoded his cosmology, Ptolemy was still held in high regard *as a scientist*. It was only in the 20th century that astronomers began to become more sceptical about some of his results: they seemed almost too good to be true . . . and in fact, on closer examination, they were. Further, it seemed improbable that he could have made some of the claimed observations at all. Once his stated results were fully analysed, it was evident that a lot of his observations would make better sense had they been made from about the latitude of the island of Rhodes. Ptolemy worked in Alexandria, but the great observational astronomer

Hipparchus (c180–125BC) had worked in Rhodes a few centuries before him. Rather than go out and make observations, it seems Ptolemy spent his time in the Library at Alexandria cribbing many of Hipparchus's results and claiming them as his own.

The clincher came when the modern researchers calculated the exact time of the autumnal equinox in the year AD132. Ptolemy recorded that he had observed it very carefully at 2pm on September 25; in fact the equinox occurred that year at 9.54am on September 24. Ptolemy was attempting to prove the accuracy of the determination Hipparchus had made of the length of the year; using as his base a record Hipparchus had made of observing the moment of equinox in 146BC, 278 years earlier, Ptolemy simply multiplied Hipparchus's figure for the year's length by 278. Unfortunately for Ptolemy's credibility, Hipparchus's figure was slightly off – hence the 28hr disparity in AD132. And it clearly wasn't just that Ptolemy adjusted the time of the observation to suit his theory; it was that he never bothered to make the observation. Had he found the equinox arrived over a day ahead of schedule he'd have been in a position to make an *even better* calculation of the year's length than Hipparchus had – and, Ptolemy being Ptolemy, this would have been an achievement he'd have crowed about.

Ptolemy wasn't the only cosmologist to cheat in this fashion. Galileo Galilei (1564–1642) certainly never conducted some of the most important experiments he reported having done when investigating gravity; those experiments would not have worked using the materials available at the time, as was evidenced by the fact that many of his contemporaries were puzzled when they couldn't replicate his results.*

John Dalton (1766–1844), famed as the propounder of the atomic theory of matter, is known to have fudged his experimental data in order to support that theory,† and even Sir Isaac Newton (1642–1727) fiddled some of the

---

* The story of his dropping weights from the Leaning Tower of Pisa seems to have been a later invention, so that's at least *one* fictitious experiment about which he didn't lie.

† An aside: Dalton was profoundly red-green colourblind, and came to believe this was because the inside of his eyeballs must contain a blue stain - a blue filter would, he reasoned, block off the red and green light.

mathematics in his *Principia* to present a more convincing case for his theory of universal gravitation.*

Another significant figure who may have faked what he's best known for is Marco Polo (1254–1324), whose ghostwritten account of his travels in the orient, *Description of the World* (*c*1298), contains some curious lacunae: there's no mention of chopsticks, foot-binding or even the Great Wall of China. At the same time, Polo did mention certain other Chinese innovations unknown to Europeans of his day, such as paper money and the burning of coal. What seems very possible is that he heard about these and other items on the grapevine via the traders with whom his family did business, and constructed the rest of his account around them. We may never know the truth of the matter. Another possibility is that his ghostwriter, Rustichello da Pisa – to whom Polo supposedly dictated his memoir while they were in prison together in Genoa – invented most of the *Description* out of whole cloth, tossing in, for the sake of verisimilitude, bits and pieces of genuine information he'd heard from Polo. Or maybe Rustichello did his best based on a somewhat incoherent narrative from Polo?†

The German physiologist Ernst Heinrich Haeckel (1834–1919) is best known today for his long-defunct Biogenetic Law – the notion that "ontogeny recapitulates

He decreed that after his death his eyes should be dissected to prove or disprove his hypothesis. The dissection was duly done, and his hypothesis disproved. Even so, red-green colourblindness is still often called Daltonism.

* As Richard S. Westfall put it in *Never at Rest: A Biography of Isaac Newton* (1980), "Having proposed exact correlation as the criterion of truth, he took care to see that exact correlation was presented, whether or not it was properly achieved. . . . [N]o one can manipulate the fudge factor so effectively as the master mathematician himself."

† We do know, though, that another account of exploration was fraudulent, even though it was given great credence in its day. This was *The Travels of John Mandeville* (*c*1371), and described the eponymous knight's journeys through much of the Near and Far East, where he encountered, so the book related, dog-headed humans, cyclopean giants, the famous vegetable lamb, and many more marvels besides. The book was not an entire fiction - the geography in it is as accurate as one could expect at the time - but clearly much of it is composed of "wonder stories"; it may have been based on oral travellers' tales. There is no clear evidence that Sir John Mandeville himself ever existed outside the anonymous author's imagination.

phylogeny" (i.e., that the physiology of the developing vertebrate embryo mimics the evolutionary development of its species). The importance of the Biogenetic Law in scientific history is frequently overstated; it has often been claimed, for example, that it played a major part in the evolutionary thinking of Charles Darwin (1809–1882), although a closer look at *On the Origin of Species* (1859) reveals that Darwin carefully made no mention of the Biogenetic Law while acknowledging Haeckel's other, valid work in vertebrate physiology. Darwin was not alone: while there's no doubt the Biogenetic Law was influential for decades, there was always sufficient public doubt for it never to be taken entirely on board. In the mid-1990s the UK embryologist Michael Richardson happened to notice that a particular diagram by Haeckel seemed curiously, well, *wrong*. This led Richardson and colleagues to conduct a more thorough examination, at the end of which they concluded Haeckel had extensively doctored his drawings of developing embryos, the main evidence supporting his Biogenetic Law, so as to fit the theory.

Louis Pasteur (1822–1895) deceived in at least two of his famous demonstrations, those concerning the vaccination of sheep against anthrax and the inoculation of humans against rabies, in 1881 and 1885 respectively. In the first instance he pretended the vaccines had been prepared by a method of his own devising when in fact he'd had to fall back on a rival method devised by a colleague, Charles Chamberland (1851–1908). The fabrication in the second instance involves the experimental work Pasteur did, or claimed he did, with dogs before his first, successful inoculation of a human against rabies. That the rabies sufferer, Joseph Meister, was cured was strictly thanks to luck, largely unsupported by any canine experiments of Pasteur's. On the other hand, without the treatment Meister would almost certainly have died anyway, so presumably Pasteur convinced himself he was taking a legitimate risk. In both instances the motive for Pasteur's dishonesty appears to have had its roots in his desire always to present himself as the master-scientist.

Sir Charles Wheatstone (1802–1875) was famed for his experiments in electricity; his name remains well known because of the Wheatstone bridge, that staple of the school science laboratory. One has to doubt whether or not Wheatstone himself had anything to do with it, going by an incident that occurred in the 1840s. The Scottish inventor Alexander Bain (1810–1877) is little remembered today, but he seems to have been possessed of the most astonishingly fertile brain, most of whose formidable powers he turned towards the then excitingly new field of electricity. Among his countless inventions were various electric clocks, telegraphy systems and railway safety systems, a way of synchronizing remote clocks, the insulating of electric cables, the earth battery and even the fax machine; he constructed a functional version of the last of these, the "electrochemical telegraph", to transmit images between the stations of Glasgow and Edinburgh.*

Developing such devices cost money, and Bain didn't have much of it. The editor of *Mechanics Magazine* was hugely impressed by Bain's work and sympathetic to his financial lack, and arranged an introduction to Wheatstone. Bain came to London in 1840 and visited the distinguished scientist, demonstrating various items, including his electric clock. The older man was dismissive: these were wonderful inventions indeed, but no more than toys; the future of electricity lay elsewhere. Very fortunately, Bain ignored this advice and applied for patents anyway, because just a few months later Wheatstone demonstrated to the Royal Society . . . the electric clock, which he claimed to have invented himself!

Wheatstone then tried to have Bain's various patents struck down; he was unsuccessful, but set up the Electric Telegraph Company, making pirated use of several of Bain's inventions. His downfall came when he sought government

---

* Bain's "fax machine" certainly worked, but in his day it could be little more than a curio: 19th-century telecommunications technology simply wasn't up to making much practical use of such a device. Over a century later the story was, of course, a bit different.

funding for the company. The House of Lords held an inquiry at which Bain appeared as a witness. Wheatstone's theft was revealed, and the company was forced to make ample financial restitution to the Scot and acknowledge him, not Wheatstone, as the true inventor of several of the devices upon which its business relied.

Of course, in the US there was no House of Lords to see justice done, and Bain's inventions were stolen wholesale. He spent the latter years of his life embroiled in interminable legal actions trying to get some measure of restitution.

Edward Jenner (1749–1823), the "Father of Vaccination", similarly claimed credit that was not his due – but got away with it. According to all the standard histories, Jenner realized that immunity to the killer disease smallpox might be conferred by inoculating people with the milder strain of the disease, cowpox, which commonly infected, but did not seriously discommode, farmers and milkmaids, well known to be less susceptible to smallpox than the general population. In 1796 he inoculated an unwitting eight-year-old boy, James Phipps, with cowpox and then a few weeks later with smallpox, revealing that the cowpox had indeed made Phipps immune to the more serious disease. This was and is generally accepted as a magnificent medical achievement. In truth it was also an astonishing lapse of medical ethics: for all Jenner knew, the experiment could have killed the lad. Or was this so? Jenner seems certainly to have been aware that over 20 years earlier, in 1774, the Dorset farmer Benjamin Jesty (d1816) had successfully performed an exactly similar action on his family – in his case not as a scientific experiment but as an act of desperation, because the county was in the midst of a smallpox epidemic. Yet Jenner never acknowledged Jesty's precedence, even while accepting a large reward from a grateful Parliament (in 1802); neither did his medical contemporaries, and almost without exception neither has history.

More recently, Alexander Fleming (1881–1955) became famed for his discovery of penicillin and thus opener of the door for the host of antibiotics that have saved so many lives around the world. In fact, although Fleming discovered penicillin – in 1928, through the accidental contamination of one of his culture dishes – and although he did name it and performed some desultory experiments with it as a possible bacteria-killer, he decided it was of little therapeutic value and moved on to other concerns. Coincidence then enters the story. A sample of penicillin was sent to the pathology department at Oxford University for use in an experiment; the stuff proved useless for that particular experiment, but was kept in case it might be handy later. A new professor took over the department, Howard Florey (1898–1968); he knew about penicillin but, like Fleming, assumed it had little medical potential. Not so Ernst Chain (1906–1979), a young Jewish refugee from Hitler's Germany. A biochemist, Chain was fascinated when he came across a reference to Fleming's discovery and investigations.

Fleming had discarded penicillin because of its instability: it was effective in killing bacteria, but only briefly. Assisted by Florey, his professor, Chain extracted a stable version, and the age of antibiotics was born. It was only then that Fleming realized the value of the discovery he had made years earlier. That discovery had been at St Mary's Hospital, London, where he worked. A governor of St Mary's, the prominent physician Lord Moran (1882–1977), was a close friend of the newspaper magnate Lord Beaverbrook (1879–1964). It would be in St Mary's interest to be associated with the antibiotic revolution, and so Beaverbrook focused the credit exclusively on Fleming – who, to his immense discredit, played along, lapping up the adulation and treating Chain's and Florey's "contribution" dismissively.* This naturally infuriated Florey, who protested to the Royal Society and the Medical Research Council; but it was not in the interests of either organization to expose Fleming's fraudulence. Some restitution came in 1945 when the Nobel Physiology or Medicine Prize went to Chain, Florey and Fleming, but even then there was an

---

* Decades later, Dr Christiaan Barnard (1922-2001) behaved similarly towards the acclaim he received as the surgeon responsible for the world's first successful heart transplant. In fact, all of Barnard's heart transplant operations were failures, as other surgeons had predicted they would be. They knew the science concerning tissue rejection was not ready for transplantation, so had held back from attempting the relatively simple operation.

injustice: anyone can make an accidental discovery; the real distinction for which the prize should be awarded is the realization of the discovery's importance and the experimental genius in making it exploitable as a life-saver.

There have been several celebrated instances in which Nobel Prizes have been awarded to one scientist for what has been mainly the work of another. Perhaps the best-known example is that of Antony Hewish (b1924), who received the 1974 Nobel Physics Prize for the discovery and early investigation of pulsars – work that had largely been done by his student Jocelyn Bell (b1943; now Jocelyn Bell Burnell). Hewish certainly supervised and guided her research, and played a significant part in the interpretation of her results, so there's no question of his not deserving the Nobel accolade, and he himself has never sought to eclipse her glory; the fault in this instance lies with the awarders.

In two other cases of the mentor–student relationship, however, the behaviour of the mentor seems more dubious. The US physicist Robert Millikan (1868–1953) received the 1923 Physics Nobel for his work establishing the electrical charge of the electron. In essence, the relevant experiment involved measuring the charge on tiny oil droplets as they fell through an electric field, and calculating from the results to reach a figure one could deduce as being the charge an individual electron must have. An experiment to attempt this using water droplets had been devised at Cambridge University but had given no very good results because the water droplets tended to evaporate almost immediately; Millikan's laboratory at the University of Chicago was trying in 1909 to refine this in hopes of getting more usable results. After months in which not much progress was made, a new arrival, graduate student Harvey Fletcher (1884–1981), had the idea of using oil droplets instead of water droplets; in Millikan's absence on other business, Fletcher set up exactly such an experiment and it worked beautifully. Thereafter the two men worked together to achieve the goal of establishing the electron's charge, and they jointly prepared a paper to this effect that was published in 1910. However, to Fletcher's dismay, Millikan cited university protocol regulations to insist the paper be published as by Millikan alone. That paper brought

Millikan the Nobel Prize; it is to Fletcher's great credit that he never expressed any ill will over the matter.

A similar controversy surrounded the award of the 1952 Physiology or Medicine Nobel to the US biochemist Selman Waksman (1888–1973) for the discovery of streptomycin, the antibiotic that effectively conquered tuberculosis. In this instance the discovery had been made, albeit under Waksman's overall direction, by his doctoral student Albert Schatz (1920–2005), and the two not only worked together thereafter on streptomycin but were listed as co-authors of the pertinent paper. Even so, the Nobel went to Waksman alone. So far, so much like the later experience of Hewish and Bell. However, Waksman then felt entitled to patent streptomycin in his name only, charging the drug companies royalties for its manufacture. These royalties were very substantial indeed, and, even though much of the money was ploughed by Waksman back into science – establishing the Waksman Institute of Microbiology at Rutgers University – Schatz not unnaturally felt he should have a share, and sued. The justice of his case was obvious: streptomycin had been a joint venture, and so the rewards likewise should be joint. This wasn't, however, how the US scientific community saw it: they were horrified any student should have the temerity to sue his mentor. All US scientific doors were firmly slammed in Schatz's face, and eventually to find work he was forced to emigrate to South America. To this day Waksman is almost always given sole credit for what is more correctly termed the Waksman–Schatz discovery of streptomycin.

Far more complicated in motivation was the public humiliation heaped by Sir Arthur Eddington (1882–1944) – during the first one-third of the 20th century the titan of astrophysics, a discipline he had almost single-handedly created – on the young mathematical physicist Subrahmanyan Chandrasekhar (1910–1995), now regarded as a far more important figure in astrophysics. In 1935 Chandrasekhar, working under Eddington, presented a paper which proved – so far as any theoretical prediction *can* prove – that stars of mass above 1.4 times that of our sun will at the end of their lives collapse infinitely. This was the first statement of the existence of black holes. What Chandrasekhar did

Sir Arthur Eddington

not know was that for some years Eddington had been working on a (largely nonsensical) Grand Universal Theory, which theory would be obviated entirely if stars could collapse into singularities. At the presentation of Chandrasekhar's paper Eddington behaved abominably, making his young protégé a laughing stock in the secure knowledge that none present would dare challenge the Father of Astrophysics. Luckily Chandrasekhar refused to take this lying down, and persisted over subsequent years in maintaining the veracity of his calculations – which were verified independently, but covertly, by such titans of the physics community as Niels Bohr (1885–1962) and Paul Dirac (1902–1984). Meanwhile Eddington used his prominence to lambast Chandrasekhar at every opportunity, framing Chandrasekhar's mathematics as incompetent while himself using every possible perversion of science – cheating, in other words – in his doomed attempt to prop up his own hypothesis. The result of all this was that Chandrasekhar's prediction of complete stellar collapse was lost to astrophysics for something like four decades. Only in 1983 was he awarded his thoroughly merited Nobel Prize for this work. Leaving his personal affront aside, the disadvantage to astrophysics through the long delay in recognizing black holes was immense – and all because of Eddington's stubbornness and the sycophancy of the rest of the astrophysics community.*

Further examples of fabrications by the great are reasonably plentiful. It seems, however, that the rate of scientific fraud has picked up remarkably since about the third quarter of the 20th century – *detected* scientific fraud,

* A highly readable account of the whole affair is Arthur I. Miller's *Empire of the Stars* (2005).

that is. The blame is almost certainly to be laid in large part on the insistence of universities and other scientific institutions – governmental science appropriations divisions not excluded – that scientists should publish prodigious numbers of papers in the learned journals as proof that they're making progress. In fact, surveys have shown that over half of such papers are never cited by other authors, and so have presumably made no contribution whatsoever to the advance of knowledge; it is suspected that many papers are in fact never read *at all* after publication. Perfectly competent, worthwhile scientists are driven to publishing unimportant papers in the more obscure and occasionally less than wholly reputable journals simply in order to impress their paymasters. The opportunities to publish fraudulent and/or plagiarized papers are, in this environment, endless.

There are other reasons for the upsurge. For example, there is very much more science being done in modern decades than in previous ones; the increase in fraud cases implies nothing one way or the other about (a) whether the percentage of fraudulent scientists is rising or (b) whether science is slowly becoming better at policing itself so that more frauds are being caught. It is disturbing, though, that the scientific establishment is not, as it claims to be as a matter of fundamental principle, even-handed in its regard for senior and junior scientists: the work of senior figures tends to go largely unchallenged while any conflicting work by junior figures is far too easily dismissed. This imbalance encourages any trends toward dishonesty among the senior figures, if nothing else through contributing to an understandable arrogance: no one can touch me. This was very evident in the case of the UK cardiologist Peter Nixon, who in 1994 sued the makers of a television documentary about his fabrication of data and his endangerment of patients; in court in 1997 he was forced to admit that the documentary was essentially accurate, something he must have known before bringing his legal action. Surely only an arrogance born out of the structure of the scientific establishment could have forced him to do something so foolish as to sue.

This weakness in the establishment's structure is even graver in the rare instances when two (or more) scientists collude to defraud. This happened during the 1990s in

Germany when the Harvard-trained molecular biologists Friedhelm Herrmann and Marion Brach enjoyed linked meteoric careers in their homeland. Because each covered the other's rear, with the pair jointly threatening any subordinate who might raise concerns about the authenticity of their work, and because no one could conceive of two charlatans being in cahoots, all might have remained undetected had the two not broken up as lovers, with the slow-growing development of the typical rancour ex-lovers can display. Once they began making acid remarks about each other's work, their juniors gained the courage to speak out. It appears the pair were fabricating data on computer to show predicted results, then writing up nonexistent experiments which they claimed had produced those data.

Brach was first to break down, admitting fakery but claiming Herrmann had put her up to it. He claimed innocence: any fakery must have been Brach's doing. Other co-authors were dragged into the melee, with at least a couple of them seemingly having followed Brach's and Herrmann's example in their own supposed work. Something like 100 papers, according to one investigative panel, had been corrupted. But then the scandal snowballed, as yet further German scientists, completely unrelated to Brach and Herrmann, started to get caught in the investigators' net. Clearly this was not just a matter of an isolated pair of crooks: there was something more fundamentally rotten in the scientific establishment of Germany – or, at least, in that part of it related to the biological sciences. Various structural changes to the set-up started to be urgently made, but it may be a while before everything is sorted out.

OF DUBIOUS HEREDITY

In 1865, just six years after the publication of Darwin's *On the Origin of Species* (1859), the Austrian monk Gregor Mendel (1822–1884) published the results of painstaking long-term experiments he had done with generations of garden peas. His concern was with the way in which characteristics were passed down from parent peas to offspring peas, and it soon became evident to him that these characteristics were embodied in the form of units that were essentially unchanging from one generation to the next: the offspring of (say) a tall and a short organism would not be all of medium height but, rather, some would be tall and some would be short. What made the offspring different from their parents was the shuffling of these various units. What Mendel had discovered was the idea of the gene, and hence the whole foundation of the modern science of heredity.

The importance of Mendel's paper was not understood until 1900, when Carl Correns (1864–1933) and Erich von Tschermak-Seysenegg (1871–1962) realized that, if this was the way inheritance worked in peas, surely so must it work in all organisms, humans included. There was a revolution in the biological sciences. Even then, biologists didn't quite grasp the tiger they had by the tail: for decades it was assumed genes were somewhat abstract entities, codings rather than actual pieces of matter which could be examined and isolated. Only with the unraveling of the structure of DNA did it come to be appreciated what a gene actually is. Today we talk blithely of splicing and engineering genes; just a few decades ago such notions would have seemed nonsensical.

Much earlier than this, in the mid-1930s, the UK statistical geneticist Sir Ronald Fisher (1890–1962) analysed Mendel's published data afresh. He found there was nothing wrong with Mendel's conclusions, and that many of his experiments could be easily duplicated to give the expected results, but that in some instances Mendel's figures were so precisely aligned to the theoretical ideal that they represented a statistical near-impossibility. There were too many spot-on results in Mendel's experiments for credibility.

What exactly happened? Did Mendel, knowing he'd uncovered a truth about the way characteristics are inherited, fudge some of his results or even invent a few extra sets of experimental results to bolster his case? Was he unconsciously biased when recording his results, seeing what he expected to see? Another possibility is that he did not in fact perform every part of every experiment himself, but had one or more assistants at work in the monastery gardens: it would be far from the first time an assistant skewed experimental results in order to make the boss happy.

This is frequently regarded, thanks to a book on the subject by Arthur Koestler (1905–1983), as the most probable explanation for the celebrated "case of the midwife toad". The Austrian biologist Paul Kammerer (1880–1926) performed a number of experiments in the first part of the 20th century which appeared to show that the Lamarckian idea of evolution by acquired characteristics was correct, rather than, or at the very least alongside, the theory of evolution through natural selection. The most dramatic of his results showed the apparent inheritance in normally land-breeding midwife toads, when forced to breed in water, of a callosity of the male palm that's found only in water-breeding toads. In water-breeding toads this callosity helps the male hold onto the female during mating; it hasn't developed in the midwife toad because, basically, mating is easier on land. For it to appear as a congenital characteristic after just a few generations of enforced water-breeding implied that offspring were inheriting characteristics their parents had acquired during the parents' lifetimes. In 1926 G.K. Noble (1894–1940) examined Kammerer's specimen toad and found the dark patch which Kammerer had claimed as the relevant callosity was in fact due to an injection of Indian ink. Disgraced, Kammerer soon afterwards committed suicide, and for a long while it was assumed he'd been a faker exposed. It's now thought at least possible his assistant "helped" the experiments.

So were the other monks "obligingly" serving up to Mendel the results he wanted? Or, feasibly, bored by the tedious pea-counting assignments he'd doled out to them, did they – unrealizing of the importance of the experiment – invent results so they could bunk off? Whatever the truth, it's a remarkable instance of a breakthrough of paramount importance being made at least partly on the basis of faked experiments.

It's a lot harder to blame an anonymous assistant in an affair that has some echoes of the Kammerer fiasco. This time the scientist at the centre of it all was William Summerlin, working in the early 1970s at, successively, Palo Alto Veterans Hospital, the University of Minnesota, and New York's Sloan–Kettering Institute. Summerlin's field was dermatology, specifically the problem of skin grafting. At the time a major problem in any attempts at grafting and transplantation was the natural inclination of the recipient's immune system to reject the grafted or implanted tissue. Summerlin was working on this, using black mice and white mice as donors and hosts so the grafted skin would show up clearly. It was while he was at the Sloan–Kettering that he was caught faking by an alert assistant, James Martin. Martin noticed something odd about the black "skin transplant" one of the white mice had received, and discovered it wasn't a skin transplant at all but had been drawn with a felt-tipped pen! In the ensuing investigation it was discovered that Summerlin's entire career of transplantation was full of dubious results, some seemingly the product of self-deception but most actually fraudulent.

In 1981–2 a similar tale emerged at the Department of Physiology at Harvard Medical School, where a young doctor called John Darsee had been researching into the efficacy of various drugs in the immediate aftermath of heart attacks. Caught faking one experiment, Darsee was permitted to continue his researches but now under close supervision. Not close enough. In the end almost all of the papers he had published during a seven-year career had to be retracted, quite a few of them on the grounds that they were complete fictions – complete with fictitious collaborators! The Darsee case became a *cause célèbre* in not just the scientific but the political world, primarily because of its poor – and slow – handling by Darsee's superiors and the scientific institutions concerned. This matter of poor, slow reaction to scientific fraud has been, alas, typical of many cases that have emerged over the past few decades.

THE IQ FRAUDS

In the field of human intelligence, two major arguments have been going on for decades. The first of these concerns IQ tests, and the extent to which they offer – or *can* offer – an accurate assessment of the individual's intelligence. The problem is that no one is certain exactly what it is that an IQ test is measuring, beyond the greater or lesser ability to do well in IQ tests. The first IQ tests were designed purely to determine whether or not certain borderline retarded children would be able to cope with school. At this the tests seemed successful. Accordingly, enthusiasts seized upon them, expanded their scope far beyond the originators' conception, and declared they were intelligence tests applicable to anyone. What got forgotten during all this was that no one had – or has – been able to decide exactly what intelligence *is*: putative definitions are countless, and none enjoys universal support.

Moreover, any IQ test will inevitably be coloured by the cultural perceptions of the person(s) who devised the test. IQ tests are superficially not tests of knowledge (or, at least, they shouldn't be), but instead tests of reasoning facility; in fact, however, they rely upon elements that, while they may be absolutely basic knowledge in one culture, may not be at all so in another. The idea that they can be applied globally with a geographic abandon has slowly withered; the idea of their applicability across cultural and/or social divides within the same geographical area is proving more persistent.

The second controversy concerns the matter of what determines the intelligence of an individual: is it a matter of heredity, or is it largely controlled by upbringing? This nature vs nurture debate seems at the moment to have swung well towards the "nurture" side of the argument, but only a bold person would claim for sure it's over. The issue is of crucial practical importance because among those insistent that the major component of intelligence is inherited is a small but vociferous coterie intent on finding "scientific" support for their racist prejudices. The irony is that xenophobia, which is what they're displaying, is largely a trait of the least developed and least intelligent.

The shrillness of some of the proponents on either side of the debate is hard to underestimate. An extreme example of an argument in favour of the "nature" side of the nature/nurture equation is "The Fallacy of Environmentalism" (1979)* by Professor H.D. Purcell (b1932). Purcell "warns" against the danger of treating all people as equal. Rather than dispassionately presenting the (defensible) case that heredity *as well as* environment plays a part in determining an individual's ability, he attacks

* In *Lying Truths* (1979), edited by Ronald Duncan and Miranda Weston-Smith.

everyone who, he thinks, disagrees with him: they are "liberals" or even, horror of horrors, "Marxists". The views of the latter aren't important because, of course, all Marxists believe in Lysenkoism!

His real concern is that researchers who come up with statistics which show that, say, one race does better than another at IQ tests are subject to unreasoned critical and sometimes even physical attack. Unfortunately, he omits to give any examples of the physical attacks to which he refers, rather in the way that modern *faux*-Christian demagogues in the US complain they're a persecuted minority while singularly failing to offer any examples of such persecution. What Purcell cheerfully ignores is that IQ tests are themselves of debatable validity, so the statistics he's so keen to defend are anyway based on at least questionable data. As A.A. Abbie (1905–1976) pointed out forcefully in *The Original Australians* (1964) on the subject of Australian Aborigines' poor results in such tests, *life itself* is an intelligence test for desert nomads: pass it or die.

The degree to which IQ tests could be put to the service of racist conclusions was exemplified by the bestselling book *The Bell Curve: Intelligence and Class Structure in American Life* (1994) by Richard Herrnstein (1930–1994) and Charles Murray (b1943). The authors believed that IQ was at least to a certain extent inherited, and pointed out a strong correlation between an individual's IQ and her or his eventual position in life: US society, they claimed, has to a large degree become stratified on the basis of IQ. A major flaw in their argument is again that IQ is certainly not a universally reliable indicator of intelligence. However, if we substitute "intelligence" for "IQ" into the authors' argument, then their point that the strata of the US's class-structured society correlate with intelligence has some limited truth (although Paris Hilton comes recurrently to mind). But it is an inescapable fact that US society is also stratified along racial lines. The authors, ingenuously or disingenuously, pointed to certain results that showed African-Americans on average doing worse at IQ tests than Asian-Americans, with white Americans somewhere in between. There are very obvious cultural reasons for this, which the authors failed to stress – and certainly many of their readers failed to appreciate the fact. The book created a *cause célèbre*, with people on one side attacking it for a racism of which the authors were, one hopes, not guilty, while people on the other side seized on what they saw as a "scientific" justification for their own racism: US society not only was but *should be* stratified along the lines of intelligence, Blacks were less intelligent than Whites, therefore it was only right and proper that Blacks should be at the bottom of the heap. If there were any logic in this at all, then Asian-Americans would be at the top of the US social heap, which manifestly they are not. At least as significant a determinant as intelligence on where the individual ends up in the social hierarchy is beauty, yet there are few who would promote the notion that the plain are inferior to the beautiful.

For decades the nature *vs* nurture debate was vastly complicated by the fraudulence of one man, a UK scientist called Sir Cyril Burt (1883–1971). Burt was a consultant psychologist to the Greater London Council in the field of education for some two decades until 1932, thereafter being Professor of Psychology at University College, London, until his retirement in 1950. During his time with the Greater London Council he was responsible for the devising and administering of intelligence tests, advising also on such matters as juvenile delinquency and the psychology of education. Not unreasonably, Burt collated the results of his IQ tests, and he became interested in the nature *vs* nurture discussion. After his retirement he returned to it, and soon produced some statistics that appeared to back conclusively his own belief that heredity was the sole significant determinant of human intelligence. He had, he claimed, located over 50 sets of identical twins who had been separated at birth and raised in different families; despite the difference in upbringing, the twins showed a very close correlation of IQs.

The matter seemed to be more or less proven, and for a long time psychologists accepted Burt's findings at face value. A year after Burt's death, however, the Princeton psychologist Leon J. Kamin (b1927) observed that Burt's figures on the identical twins seemed too good to be true; Burt's earlier average correlation of IQs between the pairs, when he had been working with only about 20 pairs, matched to the third decimal place the average he claimed later, when supposedly working with over 50 pairs. The chances of this happening in reality are statistically close to zero. Kamin also noted some vital logistic details missing from the experimental reports; for example, who had actually administered particular tests. On several occasions Burt referred to "fuller accounts" that could be found in "degree theses of the investigators mentioned in the text" – meaning, in effect, that they were unavailable, because tracking down a degree thesis is murderously difficult even in the Google era. There was a clue given in a 1943 paper of Burt's, though: the degree essay cited was "filed in the Psychological Laboratory, University College, London". Kamin investigated and discovered that, not only was there no such essay filed at the university, none had ever been submitted – indeed, there was no evidence that even its supposed author had ever existed outside Burt's imagination. In *Science and Politics of IQ* (1974) Kamin cast considerable doubt – to euphemize – on Burt's findings.

Then, in 1976, the London *Sunday Times*'s medical correspondent, Oliver Gillie, conducted some more in-depth detective work. Among the many damning new items he uncovered was that the two women Burt had cited as his research assistants, Margaret Howard and J. Conway, seemed chimerical. It became evident that Burt had very copiously faked his results – and not only in this particular study but also in others relating to the purported hereditary nature of intelligence.

This was embarrassing for others who had used Burt's findings as the basis for their own studies, among them the US psychologist Arthur Jensen (b1923), who in 1969 had published in the *Harvard Educational Review* the controversial paper "How Much Can We Boost IQ and Scholastic Achievement?", which purported to show that Black Americans scored less well in IQ tests than White Americans because they were inherently less intelligent, and not •because Black children were (as they still are) educationally deprived to a truly shocking extent. Jensen, along with his mentor, the prominent UK psychologist and author H.J. Eysenck (1916–1997) – who had himself studied under Burt – were two among an ever-dwindling few who continued to defend Burt after the revelations by Kamin and Gillie: Burt, they claimed, was far too good a statistician to have made the elementary statistical bloopers Kamin had identified, and so these must have been merely products of an old man's inattention to detail.

Burt is, by the way, far from the only psychologist to have been found guilty of fakery. The US psychologist John Broadus Watson (1878–1958), the pioneer of Behaviorism through books such as *Behavior: An Introduction to Comparative Psychology* (1914), made much of the conditioning experiments he had done with an 11-month-old infant, Little Albert. For decades these experiments were regarded as an important underpinning for Behaviorism, even though other researchers had difficulty replicating them. Only in the 1980s were psychologists willing to come out into the open and say that, while Little Albert had certainly been a real child, either Watson had never performed the groundbreaking experiments with him or, if he had, he had misreported the results.

Also of psychological interest is the concept of subliminal advertising. This seems to have been born with the report in 1957 by US market researcher James Vicary that he had exposed over 40,000 New Jersey cinemagoers to fleeting messages embedded in the movies they watched telling them to eat popcorn and drink Coca Cola; the result had been a 57% increase in popcorn sales and an 18% increase in soft-drink sales. Advertisers rushed to embed subliminal messages in their ads; contrariwise, consumer associations mounted protests. A popular and influential book on the subject was *Subliminal Seduction* (1973) by Wilson Bryan Key. However, follow-up scientific research could find no effect one way or the other from subliminal messages, and in 1962 Vicary confessed he'd made up his figures; it's doubtful he carried out the experiment. The notion of subliminal advertising was thus tossed into the litterbin where so many initially plausible scientific ideas go. Nonetheless, hopeful advertisers still apparently use the technique, if we're to judge by August Bullock's 2004 book *The Secret Sales Pitch: An Overview of Subliminal Advertising*.

THE VINLAND MAP

The curious saga of the Vinland Map began in 1957 . . . or perhaps around 1920. In 1957 the Spanish book dealer Enjo Ferrajoli and Joseph Irving Davis, of the London book dealer Davis & Orioli, brought a volume to the British Museum to have it examined. Although the volume's binding was recent, its contents seemed anything but. Its main bulk was a medieval manuscript, *The Tartar Relation* by the Franciscan monk C. de Bridia Monachi, describing an expedition by John de Plano Carpini (*c*1182–*c*1253) into Asia in 1245–7, but the real interest focused on the double-page world map, apparently of similar antiquity, bound into the front of the volume. This seemed to have been compiled around 1430–50, based on earlier Viking and other sources; unsurprisingly in this context, the coastlines of Denmark, Greenland and Iceland were depicted with a fair accuracy and the rest of Europe reasonably so, while elsewhere coastlines (and shapes) were considerably more fanciful or simply shown by wavy lines, the cartographer's indication that he was unsure of details – except for a coastline on the west of the Atlantic. On that side was shown an island whose western edge was again merely sketched but whose eastern, oceanward coast was more carefully done. It didn't take too much imagination to see in that coastline a representation of Baffin Land, Labrador and the Canadian Maritimes, with between them the Hudson Strait and the Gulf of St Lawrence. Whatever the case, here was a map dating from before Columbus's 1492 voyage, and apparently based on sources dating from long before then, that showed the New World!

The experts at the British Museum, who included the head of the BM's Map Room, Peter Skelton (1906-1970), were unconvinced – not so much on cartographic grounds as because there were mistakes in the Medieval Latin of the map's annotations. A few months later, however, the US book- and manuscript-collector Laurence Witten (1926–1995) bought the volume from its unknown owner for $3500. He showed it to his good friend the Curator of Medieval and Renaissance Literature at Yale, Thomas Marston (1905–1984), and to Yale's Curator of Maps, Alexander O. Vietor (1913–1981). The two scholars arranged that Yale should have right of first refusal should Witten ever decide to sell it. At that time, however, it was little more than an absorbing curio, since no proper provenance could be established.

That there had initially been a third volume bound in with *Tartar Relation* and the map was obvious. Both items had suffered from bookworms, yet the holes that penetrated the map did not match up directly with those of the manuscript. An inscription on the back of the map suggested the missing third item was likely to be a copy of the *Speculum Historiale*, a world history written by Vincent de Beauvais (*c*1190–*c*1264). The plain assumption was that an initial three-item volume had disintegrated with age, and that its contents had been rebound in two volumes – or perhaps the missing part had deteriorated so badly as not to be salvageable.

Some months after the meeting with Witten, Marston spotted for sale in a book catalogue issued by Davis & Orioli, Joseph Irving Davis's company, part of the *Speculum Historiale*, and bought it. This time it was his turn to show off his new purchase to Witten. Witten borrowed the volume, compared the wormholes, and a few days later excitedly announced the missing third item had been found. No one seems to have remarked too much on the astonishing coincidence involved, nor about the fact that

Marston's next action was to give his copy of the *Speculum Historiale* to Mrs Witten as a gift, to go along with the two-part volume her husband had already given to her. The bizarreness of this latter event lies in the fact that the three items together hugely increased the historical plausibility – and value – of the map. Perhaps it was an attempt by Marston to sweeten the Wittens into the idea of making sure the map came Yale's way, because of course it was now something the university very much wanted.

In 1959 the Wittens put the items up for sale, but at a price Yale couldn't afford. To the rescue came a (still) anonymous donor, who contributed nearly a million dollars. Initially Yale kept the deal under wraps, not just to allow for further authentication but also to commission a lavish coffee-table book based on the map; contributors to the book included Peter Skelton and the BM's Assistant Keeper for 15th-century books, George Painter (1914–2005). Not until 1965, with the publication of the book, did Yale announce its remarkable new acquisition to the world.

Publication day was October 11 of that year – a date chosen for maximum impact because it was the day before the US's annual Columbus Day celebration. Here, the publicists trumpeted, was proof Columbus had been not the discoverer of the New World but its rediscoverer. Of course, it had been well accepted before 1965 that the Vikings had preceded Columbus to the Americas, but that their efforts at colonization had been thwarted by a mixture of climate and native hostility. Nonetheless, the Italian–American population was outraged that "their" hero, Columbus, was being downgraded. This furore may have been one reason why, a few years later, the US declared Columbus Day an annual public holiday.

Scepticism about the authenticity of the map was not confined to gut reactions: plenty of cartographic scholars and medievalists were likewise dubious. That the matter of provenance was still hazy – Witten steadfastly refused to name the collector from whom he'd bought his volume – didn't help, but there were other concerns, mainly focusing on those errors the BM experts had noted but also on some other oddities, such as historical annotations that conflicted with general belief at the time the map had supposedly been compiled.

In 1972 Yale sent the map to the research laboratory Walter McCrone Associates for chemical microanalysis, a technology that had been far more rudimentary back in the early 1960s. The results, released in early 1974, were depressing: the map's ink contained anatase, a synthetic that wasn't invented until the early 20th century. A further analysis done in 1987, this time using particle-beam techniques, by the University of California at Davis's Crocker Nuclear Laboratory, cast some doubt on the anatase claim; however, ancillary results from Crocker supported the thesis that the map's ink was relatively recent.

There were of course two forgeries involved in the Vinland Map case: the forging of the map itself and the forging of the circumstances to convince Yale and Witten the document was genuine. The finger of suspicion for the first of these forgeries points to the Yugoslav historian Luka Jelic (1863–1922). Jelic was convinced the Vinland colonies had embraced Catholicism, and maintained that one Eirik Gnupsson (dc1121), known as Henricus, was the relevant bishop; this latter conviction was exclusive to Jelic, and yet reference to Gnupsson's putative bishopric appears in the map's annotations. It is of course entirely plausible that Jelic made the map with no fraudulent motive whatsoever, merely as an object of entertainment – he would be far from the first avid enthusiast to create a fake purely for his own amusement, rather in the same way that Elvis fans mimic their idol without any intent to deceive. The second forgery, however, is a completely different matter. As yet no clear suspects have emerged, and perhaps none ever will.

There are question marks hanging over another celebrated documentary fraud. In 1883 William Benedict (1830–1884; born Moses Shapira) tried to get the British Museum to give him £1 million for 15 parchment strips on which were inscribed the original text of *Deuteronomy*, which he said he had bought from a Bedouin who'd found them in a cave near the Dead Sea. C.D. Ginsburg (1831–1914), the BM's advisor on Semitic scripts, examined them and was impressed. What Ginsburg seems not to have known is that Benedict's track record was far from clean: the Berlin Museum had already turned down his offer, having been stung by him a decade or so earlier when he'd sold them several hundred supposedly ancient clay figures that later proved to be modern. In the interim before the BM's putative purchase, two of the strips were, with Benedict's agreement, put on public display, where they were examined by the visiting French archaeologist Charles Clairemont-Ganneau (1846–1923). He proclaimed the strips false: someone – presumably Benedict – had cut them off a Bible scroll that Benedict had sold to the BM a few years before. An alarmed Ginsburg, on re-examining the strips, agreed. Benedict reclaimed his "obvious forgery" and soon afterwards, in Rotterdam, killed himself – an act that was regarded as an admission of guilt until a few decades later when, between 1947 and 1956, the Dead Sea Scrolls came to light in caves around the Wadi Qumran. The Wadi Qumran is on the Dead Sea's northwestern shore while Benedict's Bedouin supposedly found the *Deuteronomy* parchments on the sea's eastern shore; nonetheless, the geographical proximity has raised interest in the possibility that the *Deuteronomy* parchments could have been genuine. Unfortunately, the evidence was lost after Benedict's suicide.

THE NOGUCHI CASE

The name of the Japanese–American microbiologist Hideyo Noguchi (1876–1928) may be obscure now, but in his day he was something of a celebrity in his field. He was brought to the US by Simon Flexner (1863–1946), renowned for having isolated the organism responsible for dysentery. Noguchi's research followed along similar lines . . . only even more so: he managed to isolate polio, rabies, syphilis, trachoma and yellow fever – an astonishing tally for a single scientist! After Noguchi's death – of yellow fever – pathologist Theobald Smith (1859–1934) spoke for many when he remarked: "He will stand out more and more clearly as one of the greatest, if not the greatest, figure in microbiology since Pasteur and Koch." Unfortunately for Noguchi's reputation, research done over the next few decades slowly revealed the truth: he had isolated none of these diseases, and much of the rest of his work was likewise specious. Whether he was deliberately fraudulent or whether he was simply guilty of self-deception through overeagerness, no one knows.

THE SPRAGUE/BROWNING CASE

One consistent problem about exposing fraudulent scientists is that the people who blow the whistle on the fraud are often the first to be punished by the scientific establishment, and the more severely so, while the perpetrator of the fraud is not uncommonly given the benefit of the doubt for years, or even forever. In short, the messenger is shot. A typical example is that of the University of Illinois psychologist Robert L. Sprague, who blew the whistle on the fabrication of data by Stephen Breuning. When Sprague first met Breuning he was much impressed by the young graduate student, who, working at a center for the mentally retarded, was in a position to help Sprague's researches into neuroleptic drugs, used in the treatment of such patients. The two men collaborated happily for a couple of years, until 1981, but then Sprague became concerned that Breuning's results were *too good*: there didn't seem to be enough hours in the day or days in the year for Breuning to have conducted all the studies he reported. The final straw was Breuning's claim that subjective assessments of the patients made by nurses showed 100% agreement; in practice, one's lucky – or self-deluded – to obtain as much as 80% agreement.

Sprague, rightly, blew the whistle on Breuning, by now working at the University of Pittsburgh's Western Psychiatric Unit. This was not an unimportant issue, because Breuning's "results" showed strongly that neuroleptics were often doing patients more harm than good; in extreme cases, taking a patient off a course of the drugs could result in an assessed doubling of the patient's IQ! In response to publication of Breuning's papers on the subject, physicians worldwide were becoming reluctant to use neuroleptics. The University of Pittsburgh set up a committee to examine Sprague's claims but, although Breuning confessed an earlier fabrication and resigned, the committee declined to examine any of the equally fraudulent work he'd done at their own university: so far as it was concerned, Breuning was guilty of no more than a youthful indiscretion.

Appallingly, not only did the National Institute of Mental Health accept this coverup, it began to investigate

Sprague. His federal grants were cut off. When in 1988 he testified about the case in front of Congress's Subcommittee on Oversight and Investigations, the first reaction of the University of Pittsburgh was to threaten him with a libel suit. Only when the Subcommittee's chairman, John Dingell (b1926), stepped in did the university climb down and produce a grudging letter of apology. By then, of course, Sprague's career had suffered untold damage.

If this were an isolated example, matters would be bad enough. As we shall see, it isn't.

THE CANTEKIN/BLUESTONE CASE

A few decades ago, children's earaches were generally treated at home: they were regarded as something transient that would, if left alone, in due course clear up. Gradually, however, it became apparent that in at least a few instances earaches could be signs of something more serious: ear infections might lead on to deafness, meningitis or even, on rare occasion, death. With the advent of antibiotics it seemed that at last there might be a quick, simple, easily available remedy. Research turned toward finding out if this was indeed the case; and, if so, to determine which antibiotics should be used.

In the early 1980s Drs Charles D. Bluestone and Erdem I. Cantekin, working at the Otitis Media Research Center of the University of Pittsburgh, set about a large-scale double-blind clinical trial of an antibiotic called amoxicillin. They received a fair amount of government financial help via the National Institutes of Health, but seemingly not enough, because Bluestone – despite Cantekin's forebodings – persuaded a number of pharmaceutical companies to provide additional funding, not to mention over $250,000 in honoraria and travel expenses for himself; the companies involved included Ross Laboratories, Eli Lilly and Beechams. Further, again overriding Cantekin's objections, he redesigned the trial to include alongside amoxicillin two commercially produced and considerably more expensive variants: Ceclor, manufactured by Eli Lilly, and Pediazole, which came from Ross.

By this time Cantekin was significantly worried by the

way things were going. Analysis of statistical data is a troublesome business at the best of times, yet it seemed clear to him that the interim results of the trials of amoxicillin showed no distinct advantage in using the antibiotic over the placebo that had been administered to other patients among the trial group. Bluestone's conclusion differed: he saw amoxicillin as having a definite edge. Part of the problem was that the two men had different ideas on the moment at which a cure could be said to have been effected: Bluestone considered that, if the ear were still clear of infection after four weeks, that could be checked off on the list as a cure, while Cantekin, pointing out that ear infections often came back for a second round, insisted eight weeks was a better yardstick. Using Bluestone's rule, amoxicillin performed marginally better than the placebo, but using Cantekin's there was no perceptible difference. And, if amoxicillin were ineffective, what was the point of comparing the performance of Ceclor and Pediazole against it?

In 1985 Bluestone and Cantekin considered the study complete, and set about producing their report for publication. Soon Cantekin found himself appalled by the conclusions reached by Bluestone and the paper's other co-authors, and withdrew from further proceedings. The finished paper was submitted to the *New England Journal of Medicine* . . . but so too was an alternative paper, written by Cantekin alone and, though based on exactly the same research, coming to an opposite conclusion: there was no clear evidence that antibiotics, specifically amoxicillin, were effective against ear infections. The journal's editors, understandably confused, approached the authorities at the University of Pittsburgh to ask which of the two was the "genuine" paper; those authorities responded that, since Bluestone was team leader, his paper was the official one – and so the journal accepted it and rejected Cantekin's rival version.

Some while later the *New England Journal of Medicine* gave its reason for the decision: "The important question . . . is not whose interpretation is correct . . . but who has the right to publish." One reader of this retorted:

> Let me reword your statement just slightly to demonstrate its astonishing absurdity: "The important question . . . is not

whether Galileo or the Pope is correct with respect to whether the sun revolves around the earth or vice versa, but rather who, as between the Pope and Galileo, has the right to publish his opinion.[*]

Cantekin did not take any of this lying down. He felt Bluestone's conclusions had been hopelessly contaminated by the commercial funding. Further, and more important, he was still unconvinced that antibiotics were helpful in ear infections. The importance of this goes beyond the matter of patients wasting money on – and companies reaping profits from – a drug that doesn't work. In a few instances patients would suffer grossly from their ear infections having been left, in effect, untreated. Even more worrisome was the known fact that the gratuitous use of antibiotics can result in the generation of strains of "superbugs".

For five years Cantekin waged his struggle unsuccessfully, accusing Bluestone of perpetrating a fraud. Various panels at the University of Pittsburgh and the National Institutes of Health rejected his claims. One NIH report did suggest Bluestone might indeed have been – perhaps quite unconsciously – guilty of tilting his conclusions in favour of his funders, and recommended his further researches be subjected to a five-year period of "administrative oversight", but that was it.

At least, that was it until 1990. In that year Congress's Subcommittee on Human Resources and Intergovernmental Relations held hearings on scientific misconduct that could endanger the public health, and the Cantekin–Bluestone affair was one of the cases they considered. The Subcommittee came down like a ton of bricks on the University of Pittsburgh and the NIH, especially for the fact that between them these two institutions had managed to suppress Cantekin's conclusions almost entirely: "Evidence of the ineffectiveness of antibiotics would have been available to physicians and the public several years ago if the medical school had not prevented Dr Cantekin from publishing [his conclusions]."

The subcommittee's declaration may well have been coloured by the fact that, in what could be considered the

* Both the *NEJoM*'s comment and the response are as cited in Michael O'Donnell's *Medicine's Strangest Cases* (2002).

largest experimental trial of all – the administration by doctors of antibiotics to millions of patients suffering ear infections – similar conclusions to Cantekin's were being reached. As Cynthia Crossen reported in her valuable account of the whole affair in the *Wall Street Journal* (January 3 2001):

> . . . as the dispute has moved slowly through these tribunals, medical science has gradually come to its own conclusions about antibiotics and ear infections – and they are in line with Dr Cantekin's. Although more antibiotics are prescribed today for children's ear infections – and for longer periods of time – in the US than anywhere in the world, several recent, independently financed studies have found that, for the vast majority of ear infections, antibiotics are little more effective than no treatment at all. Worse, physicians are now seeing in their own practices the potentially deadly consequences of too many children taking too many antibiotics – drug-resistant strains of bacteria.

In other words, Cantekin had been exactly right.

One might have thought the matter would be over after Congress's very public exoneration of Cantekin, but no. In 1991 Bluestone published a further paper based on the 1980s data, again concluding that amoxicillin "or an equivalent antimicrobial drug" was the treatment of choice for ear infections even though the data themselves showed at best a borderline effect. These conclusions became part of a federal Clinical Practice Guideline. And so Cantekin finally turned to the law . . . and years passed. Not until 1999 did the legal tide begin to turn in Cantekin's favour. In the meantime he had been, ever since the whole affair blew up, completely ostracized by his colleagues at the University of Pittsburgh – not fired, presumably because that would have brought the matter even more emphatically into the public view – but with his salary frozen at 1986 levels. Yet again, the honest whistleblower was punished for acting in the interests of the public welfare.

THE VARDY/FRENCH/MCBRIDE CASE

The Australian obstetrician and gynaecologist William McBride (b1927) was the man who in a 1961 letter to *Lancet* fingered the drug Thalidomide, often prescribed for pregnant women, as the cause of so many babies being born with shrunken limbs. With good reason, he became an international celebrity. In 1981 history looked set to repeat itself when he and two junior colleagues, Phil Vardy and Jill French, published a paper raising the alarm about another drug commonly prescribed for pregnant women, Debendox (marketed in the US as Bendectin): the three scientists had administered the drug to pregnant rabbits, some of whose offspring had been born with, once more, deformed limbs.

The only trouble was that, when Vardy and French read the published version of the paper, they immediately saw their data had been significantly manipulated – aside from anything else, a couple of fictitious pregnant rabbits had been added to the roster. The two young scientists complained to McBride. They were promptly sacked, and locally blacklisted. They tried to take the matter to higher authority, but their petitions were rejected out of hand, as was the letter they wrote to the journal in which the falsified paper had appeared, the *Australian Journal of Biological Sciences*.

Meanwhile, the lawsuits began, about 1300 of them: any parents whose children were born with abnormalities where the mother had taken Debendox not unnaturally went to court about it. The drug's makers, Marion Merrell Dow, found many of the cases going against them, not least because McBride was only too willing to appear as an expert witness; fortunately for justice, Marion Merrell Dow appealed in all cases, the company's own clinical tests having found nothing wrong with the drug. Even so, the lie might have remained undetected in perpetuity had the Australian medical broadcaster Norman Swan not managed, fully five years later, to worm the story out of a reluctant Vardy. Swan's broadcast, and a subsequent book by Bill Nicol, *McBride: Behind the Myth* (1989), clinched the case, yet still the Australian medical establishment obstructed any attempts to take action. Finally, though, the

New South Wales Medical Tribunal was forced to hold a hearing, and at the end of lengthy deliberations found McBride guilty of fraud. This, of course, did nothing to compensate Vardy and French for the destruction of their careers, and nor did it do much for all those families that had invested everything in honestly mounted lawsuits against Marion Merrell Dow and now discovered themselves confronted by huge legal bills.

QUESTIONS OF AUTHORSHIP

In 1995 the UK gynaecologist Malcolm Pearce was struck off the medical register by the General Medical Council. In the previous year he had published in a single issue of the *British Journal of Obstetrics and Gynaecology*, a journal of which he was an editor, two papers concerning research which he had in fact never done. His co-authors – two on one paper, one on the other – had had nothing to do with the papers, having accepted their "co-authorships" as a "gift"; though two of the co-authors were relatively junior, the third, Geoffrey Chamberlain (b1930), was the journal's Editor-in-Chief as well as President of the Royal College of Obstetricians and Gynaecologists and Pearce's departmental head at St George's Hospital, London. The two had also cowritten a book, *Lecture Notes in Obstetrics* (1992). In the paper upon which Chamberlain's name was listed, Pearce claimed to have treated an ectopic pregnancy (one in which the embryo has implanted outside the uterus) with such success that both mother and baby were now doing fine. This would have been an historic feat, if true. The second paper concerned a large-scale trial on a rare ailment which would have involved immense logistical outlay and yet on which none of Pearce's colleagues could remember him working.

Chamberlain and the two junior "co-authors" were reprimanded, and to be fair it was Chamberlain who initiated the probe once people had started muttering about the papers; further, Chamberlain stepped down from his editorial job and from his RCOG presidency. It seems to have been assumed by all concerned that the two papers were a rush of blood to Pearce's head, and that all his earlier work was still valid; yet experience of other fraudsters tells us this is unlikely. Other disturbing aspects include the nepotism of editors publishing their own papers and the matter of academics accepting prestigious "co-authorships" of papers on work of which they in fact knew nothing; it's a practice which, while common,* is dishonest on numerous levels.

In 1978 Helena Wachslicht-Rodbard of the National Institute of Arthritis and Metabolic Diseases submitted a paper on anorexia nervosa (further details need not concern us) to the *New England Journal of Medicine*. It was sent out in the ordinary way to two specialists in the field for peer review; one of these was Philip Felig (b1936), Vice-Chairman of Yale's Department of Medicine. He gave the paper to a junior colleague, Vijay Soman, to do the actual reviewing. Soman's opinion, sent to the journal's editor as Felig's, was negative. The other review was positive, as was that of a third specialist to whom the editor then sent the paper as umpire. The whole process took three months, and in the interim Soman and Felig submitted to the *American Journal of Medicine* a paper, based on work by Soman, that bore an astonishing resemblance to Helena Wachslicht-Rodbard's original. Indeed, as she discovered when, in a delightful symmetry, the *AJM*'s editor sent the paper to her boss Jesse Roth for peer review and Roth passed it on to her to do the actual reviewing, bits of Soman's paper were actually copied verbatim from her own. In due course – a very long due course – it was proven that Soman, who'd been planning a similar study to Wachslicht-Rodbard's, had saved himself a bit of work by simply plagiarizing her material. Soman's career was, quite reasonably, destroyed. Felig, who had been a powerful man in the field, was deemed to have behaved badly throughout the affair and suffered too, being forced to resign the position he'd just taken up as head of the Department of Medicine at Columbia University's College of Physicians and Surgeons.

Another plagiarism case that surfaced at about the same time was that of Elias A.K. Alsabti (if that was indeed his real name), who was born in Basra, Iraq, but was, by the time he came to the US in 1977, supposedly a citizen of Jordan, where he had, he said, gained an MB and a ChB

---

* And not just in the field of academic papers but in the world of commercial publishing, too.

(Bachelor of Surgery). He claimed a blood relationship with the Jordanian Royal Family, who had paid for him to come to the US to gain his PhD. Alsabti charmed his way into posts at various non-negligible institutions across the US and published prolifically: some 60 papers over a period of about three years. He was able to publish so much because his technique was simple: he would take a paper from a little-known journal, copy it, alter the title, claim the paper's authorship (with or without invented co-authors), and send it off to *another* little-known journal in a country different from the first. Since the vast majority of published scientific papers go largely unread, Alsabti might have continued this practice for very much longer than he did. He was just plain unlucky to get caught: a student of microbiologist E. Frederick Wheelock, of the Jefferson Medical College, Philadelphia, spotted that a paper published in Czechoslovakia by Alsabti was almost identical to a paper, also "authored" by Alsabti, that had appeared obscurely in the US . . . and that both were just redigested versions of papers Alsabti had stolen from Wheelock on being fired from Jefferson for basic incompetence. Wheelock whistle-blew, writing to the editors of all the major journals in the field; but, with the exception of *Lancet*, all shamefully declined to publish his letter on the grounds that this was just a personal quarrel between the two men. As a result Alsabti's US career as a conman was allowed to continue for several more years before unearthed examples of his plagiarism became too many for the scientific establishment to ignore.

Alsabti was fortunate indeed not to have been caught earlier, in Iraq, where in 1975 he persuaded the government of Ahmed Hassan Al-Bakr that he had invented a brand-new cancer diagnostic technique. Soon he was running his own lab in Baghdad, while also visiting local factories and fraudulently charging the workers there a fee for screening them for cancer. Since the Iraqi Government offered universal health care, Alsabti's fee-charging habits eventually came to the attention of the authorities, but by the time they moved in he'd fled the country, eventually ending up in Jordan, where he claimed political asylum and, in essence, pulled the cancer stunt all over again – to the extent that the Jordanian Government did indeed finance his transition to the US. Ironically, the Jordanians were among the first to recognize his fraudulence in the US, cutting off his funding when they spotted he was listing senior Jordanian scientists as his co-authors without their knowledge. Meanwhile, even after Wheelock sounded the alarm, the US scientific establishment was letting Alsabti happily loot onward . . .

Action was much swifter in the case of the Polish chemical engineer Andrzej Jendryczko, whose plagiarism was first revealed in 1997, the scope of it becoming more evident over the next few months. His method was to take articles from primarily English-language scientific journals, translate them into Polish, then publish them as his own. Since comparatively few people outside Poland read the lesser Polish journals, he must have thought he was safe enough. Perhaps as many as 100 plagiarized papers, perhaps even more, were published under his name, the names of various co-authors being added with or without their knowledge. It seems there was a brief attempt at a coverup by the Medical University of Silesia, where Jendryczko had worked, but it was a matter of mere weeks before the authorities there made a public admission.

The creation of false data lay at the heart of a US scandal of the 1990s, although the true scandal here was that the data had been fabricated not during the 1990s – that was merely when the crime came to light – but in the 1980s. The perpetrator was Roger Poisson, a researcher at St Luc's Hospital, Montreal. He was a major contributor to clinical trials being conducted under the aegis of the US National Surgical Adjuvant Breast and Bowel Project, with Bernard Fisher (b1918) of the University of Pittsburgh as the primary organizer. The point of the trials was to find out if, in many cases of breast cancer, lumpectomy – excision of the tumour, followed by other treatment – could offer just as great a chance of success as the then-prevalent option of mastectomy. To the great relief of many women, the conclusion of the trials, presented by Fisher and Poisson from 1985 onwards, was that lumpectomy was indeed a perfectly viable procedure in many instances.

In 1990 Fisher's team grew suspicious of Poisson's data. In 1991 Fisher, having reevaluated much of Poisson's work, reported the matter to the National Cancer Institute. The

NCI passed things on to the Office of Scientific Integrity, who took until 1993 to issue a definitive report that Poisson had indeed been doctoring his data. Not until March 1994, when the story broke in the *Chicago Tribune*, was the public made aware of what had been going on – yet surely this was a topic of crucial interest to said public. Further, in the period from 1990 to 1993 Fisher and the National Cancer Institute had done nothing to alert the scientific community to the fact that papers involving Poisson must be regarded as suspect. Fisher's defence was that, in re-evaluating the trials while excluding Poisson's contribution, there was no reason to change the conclusions.* That does not excuse the coverup.

CURES FOR CANCER

Of all the quacks, perhaps the most contemptible are those who peddle false cures for cancer. The deaths they have caused, through diverting their victims from therapies that might at the very least have offered a chance for survival, are a matter of record, and yet these individuals, at least in the US, have shown the tenacity of a nasty yeast infection, using the fallibilities of the law to extend, often by decades, their lethal and avaricious careers.

It was a case of cancer that succeeded finally in bringing to an end the career of Ruth B. Drown (1891–1965), a quack who made a lifetime career using her remarkable Radio Therapeutic Instrument.† Mrs Marguerite Rice of Illinois was told by her doctor in 1948 that a lump in her breast might signify cancer; he urged her to have a biopsy taken. But Mrs Rice was frightened of the procedure, so she and her husband

---

* And indeed those conclusions still stand today.
† A far more detailed account than is possible here of both this case and the succeeding one can be found in James Harvey Young's *The Medical Messiahs* (2nd edn 1992).

instead consulted Drown. Drown referred the Rices to a local practitioner, Dr Findley D. John, and Mrs Rice had several weeks' worth of expensive sessions with him before Mr Rice decided it might be cheaper to invest in their own personal Radio Therapeutic Instrument. That bought – again not cheaply – Mrs Rice spent hours each day "completing the circuit". Some while later the Rices came across a newspaper article debunking Drown's therapy. They rapidly sought the medical attention they should have sought in the first place, and at the same time contacted the American Medical Association. The AMA in turn advised the Food and Drug Administration (FDA), who had for some time been looking for a means of prosecuting Drown but had had their hands tied by legal niceties. In the event it was not until 1951 that the FDA was finally able, bolstered by the Rices' evidence, to bring Drown to trial, in Los Angeles, in her home state of California – although by this time Mrs Rice, her cancer inoperable, was too weak to attend. Drown was found guilty, and fined $1000. Even after this, she could still continue to practise in California so long as she did not sell her devices across state lines. So on she carried, with continued financial success. Only in 1963 did the California State Bureau bring a case against her, and in the event Drown died before that case could be brought to trial. Mrs Rice, of course, had died long since.

Even more tortuous and even more alarming was the case of Harry Mathias Hoxsey (1901–1974). Hoxsey claimed to have inherited his cancer cure from his father, who had died – of cancer! – in 1919; the father supposedly had observed how a horse had cured itself of a leg cancer by standing in a patch of certain shrubs, and had gone on from there to treating first animals and then human patients. He had passed the secret to young Harry on his deathbed.[*] However much truth there is in this story, it was in about 1922 that Hoxsey first used the compound, a paste, smearing it onto the lip cancer of a Civil War veteran and apparently curing him. It is indeed possible that the man was cured, because some years later, when the AMA was able to get hold of a sample of Hoxsey's paste, they discovered its active ingredient was arsenic. Arsenic is in this context an

* Hoxsey told various versions of the tale, but this is the gist of it.

escharotic: it corrodes away surrounding cells, including cancer cells. That it also destroys healthy cells may or may not be important, depending on the circumstances; more usually it is, and the dangers become even more extreme should the arsenic get into the bloodstream, as any mystery-fiction fan knows.

In 1924 Hoxsey opened a practice in Taylorville, Illinois, at the same time co-founding the National Cancer Research Institute – a grand-sounding name for a commercial operation designed to market Hoxsey's cure. The other co-founders soon pulled out, so Hoxsey opened instead the Hoxide Institute, advertising for patients. Soon the local newspapers were running articles about deaths at the Institute, while the *Journal of the American Medical Association* took up the cudgels concerning Hoxsey's fraudulence. Hoxsey responded by suing the AMA, a case that was dismissed when the court wearied of the Hoxide Institute's endless delays. In a separate case, however, Hoxsey was convicted of practising medicine without a licence and fined $100, and in 1928 the Hoxide Institute closed.

At least in Taylorville. After a brief attempt to reopen for business in Jacksonville, Illinois, Hoxsey returned to his hometown of Girard, Illinois, with more success – although Arthur J. Cramp, a contemporary scourge of quacks, commented dourly: "Perhaps Girard will flourish briefly – especially the local undertaker and those individuals who have rooms to rent." However, as in Jacksonville, Hoxsey's Girard venture petered out swiftly, and the story was repeated in several other states. In 1936 he reached Dallas, Texas, where he set up in grander style than ever before. He was convicted of practising without a licence yet again, and this time the punishment – a five-month jail sentence an $25,000 fine – was more than a slap on the wrist; unfortunately, the verdict was overturned by a higher court. He then managed to obtain a Texas licence to practise naturopathy, which meant it was difficult for the authorities further to pursue him.

By now he was going beyond the treatment of external cancers, prescribing "chemicals" for the treatment of internal tumours. The medical hypothesis he put forward to justify the use of his medicines was that cancer was a result of chemical imbalance within the body; his chemicals supposedly restored the balance, thereby checking the cancer and driving it out.

Visitors to his clinic were subjected to an impressive battery of diagnostic tests, the results of which indicated in a surprisingly high number of cases that the patient was indeed suffering from cancer. Later, when being challenged in and out of court, Hoxsey and his startlingly incompetent Medical Director, one J.B. Durkee, were able to claim impressively high cure rates (for the time) precisely because so many of their "successfully cured" patients proved, on proper examination, never to have been suffering from cancer in the first place. But Hoxsey and Durkee were committing a far greater crime than fleecing the gullible for treatments for a disease they did not have: the two men were also actively discouraging *genuine* cancer sufferers from pursuing orthodox treatments, claiming these not only didn't work but actually made the condition worse.

To the modern mind it seems completely incomprehensible that Hoxsey and his cronies could have been permitted to continue in their careers of destroying lives, but this is to underestimate the gravity of the matter. Hoxsey played politics astutely to become a significant figure, drawing prominent Senators into his web. Later on he was allied for a while with the pro-Nazi Christian evangelist Gerald B. Winrod (1900–1957), upon whom Sinclair Lewis based his fascist demagogue Buzz Windrip in the novel *It Can't Happen Here* (1935).* Winrod accepted large sums of money from Hoxsey to plug the latter's fraudulent cures in Winrod's racist and antisemitic newspaper *The Defender*. Winrod's was not the only fascist organization Hoxsey flirted with.

Despite his connections in high places, legal proceedings dogged him. Some were self-inflicted. In 1947 Morris Fishbein described Hoxsey in the *Journal of the American Medical Association* as a "cancer charlatan", and Hoxsey sued him for libel. He won the case, but was awarded not the million dollars he asked for but only a dollar apiece for

---

* A dire warning in the vein of Orwell's *1984* (1949), showing how the US could slip into fascist totalitarianism.

himself and his dead father, whom he claimed had also been libelled.

The judge in that case, William Hawley Atwell (1869–1961), was rumoured himself to have been one of Hoxsey's "satisfied customers", and so should have recused himself. Neither did he recuse himself when the FDA brought their major case against Hoxsey, in 1950, accusing him of shipping fraudulent medicines across state lines; Atwell's conduct of the proceedings in court was so manifestly biased in Hoxsey's favour that it came as no surprise to the FDA that they lost the case. A prompt appeal to a higher court reversed the verdict, and Atwell was instructed to issue an injunction. This he did, but in such a watered-down form that it was of no effective worth. Only on a second attempt was Atwell compelled to issue the injunction the FDA had initially sought.*

But Hoxsey was still free to continue his practice in Dallas! He published a ghostwritten autobiography, *You Don't Have to Die* (1956), which sold well and attracted further customers. And, in James Harvey Young's memorable phrase, "Hoxsey's cancer treatment metastasized from Texas into other states." All this time the FDA was powerless to do anything, because, incredibly, Hoxsey's activities were within the law.

The FDA was, however, entitled to issue public warnings about dangerous drugs, and in 1956 it began a campaign against him. Hoxsey brought a lawsuit against the FDA, but lost. Finally emboldened, the Texas legal authorities began to take some action of their own. Hoxsey did not give in without a struggle, and there were various further legal manoeuvres, but by about 1960 the cancer of the Hoxsey cure had been more or less stamped out.

Well, not entirely. As late as 1987 the feature-movie documentary *Hoxsey: How Healing Becomes a Crime* sought in part to rehabilitate his reputation, using the kind of "balance" that bedevils the modern US news media.

While supercures for cancer have been almost the exclusive province of the quack, there have been a few

---

* Atwell found further ignominy when in 1957 he again had to be forced by higher courts to abide by the law, this time the law concerning school desegregation.

touters of miracle cancer cures who have, at least for a while, seemed perfectly respectable – and sometimes even eminent. One such was Abdul Fatah al-Sayyab, a scientist who in the early 1970s rose to a position of considerable prominence in the Iraqi government thanks to the two drugs he had derived, tactfully called Bakrin and Saddamine in honour of the President and Vice-President of Iraq at the time, Ahmed Hassan al-Bakr (1914–1982) and Saddam Hussein (1937–2006). According to al-Sayyab, these drugs could cure certain types of cancer – hence his rise to high political status. In due course, alas, it was discovered the drugs had no effect on cancerous growth whatsoever, and al-Sayyab fell from grace.

The early 1980s saw a much more serious claim to have gotten a handle on cancer, thanks to Professor Efraim Racker (1913–1991) of Cornell University and his graduate student Mark Spector (bc1957). In consequence of a remarkable series of experiments performed by Spector, they put forward a new theory, the kinase-cascade theory, to explain the cause of at least some forms of cancer. The theory is slightly complex for non-biochemists, but in essence Racker suspected that malfunctioning of a particular enzyme present in the walls of all cells, sodium–potassium ATPase, was symptomatic of cancer cells. Spector's experiments not only confirmed this but went further: the malfunctioning of the ATPase was due to the action of another enzyme present in all cells, a protein-kinase. Harmless in normal cells, the protein-kinase became actively harmful in cells that turned cancerous. More experiments showed that it, too, was being triggered by an enzyme – another kinase, in fact. All told, Spector reported, there were four kinases involved, acting in cascade fashion each to modify the next, with the last modifying the ATPase.

Spector then tied this model in with a recent discovery concerning the genes of viruses that cause cancer tumours in animals. The gene responsible, the so-called sarc gene, programs the virus to produce a protein-kinase. No one had been able to find a sarc gene in any organism other than a virus; at the time the only way to look for the gene was to search for its product, the protein-kinase, and this search had been in vain. Not when super-experimenter Spector turned to the task, however. He was soon able to show that some of the kinases he'd isolated were sarc-gene products.

A model for cancer's mechanism now seemed within reach. When a tumour virus invaded a cell, its sarc genes triggered a kinase cascade which ended in the adverse modification of the ATPase, so that in due course the cell turned malignant. If the mechanism of cancer could be understood, surely a cure or at least a preventative could not be far behind.

The trouble was that no one could replicate Spector's experiments – unless he happened to be there to help. Rather embarrassingly, not many tried; further, a couple of scientists who became suspicious of the necessity for Spector's presence kept their suspicions to themselves. The bubble was finally burst by Spector's Cornell colleague Volker Vogt, who decided to investigate for himself the gels Spector was using to provide radiometric proof of the presence of his kinases. Vogt discovered that a very clever, sophisticated trick was being used with the gels to produce the results Spector wanted – indeed, the whole model for cancer's mechanism proved to have been based on Spector's faked results. It was about now that Cornell decided to check up on its star graduate student: it emerged he had neither of the two earlier degrees he claimed but did have a couple of charges for forgery on his record.

No one knows what Spector's motives were for faking the kinase experiments, although the quest for scientific glory must have been part of it – there was much talk of a Nobel Prize for himself and Racker. As for Racker, it seems he was so keen for his hypothesis concerning ATPase to be shown correct that he was in a state of denial as to the need properly to double-check his younger colleague's experimental results – or he may just have been bamboozled by a charming conman, as can happen to anyone: science often relies, after all, on the assumption that one can trust the integrity of fellow-scientists. The main consequence of the whole hoax was, of course, far more important than the vast amount of time and energy spent on the kinase-cascade theory: cancer research was diverted down a blind alley for two long years or more, while the hopes of some of the sickest people on the planet were raised to a high pitch, only to be cruelly dashed.

THE O'TOOLE/IMANISHI-KARI/
BALTIMORE CASE

What has come to be known as the Baltimore Affair really concerns the immunologist Thereza Imanishi-Kari (b1943); the focus came to be on the role of David Baltimore (b1938), one of the world's foremost molecular biologists, because of his intransigence as he attempted to cover up Imanishi-Kari's misdeeds and destroy all dissenters – in particular the young researcher, Margot O'Toole, who had blown the whistle on Imanishi-Kari.[*]

O'Toole had come across evidence that Imanishi-Kari had falsified results in at least one of the experiments she had conducted – an experiment that was fundamental to the work Imanishi-Kari's lab at MIT (it was just about to move to Tufts University Medical School, Boston) had been doing on behalf of David Baltimore's much larger lab at the Whitehead Institute. O'Toole was particularly well qualified to recognize the evidence because the experiment concerned was one that had led to a falling-out between herself and her boss, Imanishi-Kari: O'Toole had, working with Imanishi-Kari's colleague Moema Reis, obtained a result that seemed supportive of Imanishi-Kari's work, but had been unable to replicate the result when repeating the experiment on her own and so had, to Imanishi-Kari's fury, declined to publish the finding. Now she had uncovered proof that Imanishi-Kari herself had repeatedly gained the same negative result. If Imanishi-Kari was prepared to falsify one result in order to bolster the hypothesis Baltimore was currently proposing, in how many other instances might she have done the same?[†] Imanishi-Kari and Baltimore had collaborated on a whole series of research papers that all pointed in the same direction, the latest of which (universally known as *Weaver et al.*) had been

---

[*] An extensive discussion of the case appears in Horace Freeland Judson's *The Great Betrayal* (2004).
[†] The research concerned the behaviour of transplanted genes and their associated proteins. It seemed very important at the time, since it would have been a clincher for one of the two sides in a major debate that was then much occupying the molecular-biology world. In fact, the side it was supporting has since been proven wrong.

received as of great significance. But if Imanishi-Kari had merely been dishing up to Baltimore the results that he wanted . . .

O'Toole did not immediately whistleblow; she did, however, say that *Weaver et al.* should be retracted pending further investigation of its subject matter. She did all the right things, taking her concerns privately to senior figures at MIT and Tufts; but here the familiar pattern emerged of the establishment putting every possible obstacle in the whistleblower's path. There was a very extensive coverup, during which administrators and senior scientists collaborated disgracefully to pretend there was nothing untoward and to drum O'Toole out of her research career.

O'Toole tried to put it all behind her, and behind her it would have remained had it not been for, once again, Imanishi-Kari's abrasive personality. A former graduate student in her lab, Charles Maplethorpe, had kept in touch with his old acquaintances there, O'Toole included. Disturbed by what was going on, and having himself had doubts during his time at the lab about the veracity of some of the research, he contacted Walter W. Stewart and Ned Feder, the two self-appointed "policemen of science" who some while earlier had exposed the Darsee *imbroglio* (see page 33). It was through Stewart and Feder that the case was brought before John Dingell's Subcommittee on Oversight and Investigations in 1988–9. Baltimore was one of the witnesses called, and he used the public opportunity to charge that Dingell and his colleagues were conducting not a reasonable investigation of a specific case but more generally a witchhunt against scientists and scientific research everywhere.

Baltimore had enormous standing and the motives of politicians when interacting with science are often corrupt, so it's hardly surprising that many scientists across the US took Baltimore's statement at face value without pausing to examine what was actually going on. A national movement swiftly grew . . . and as swiftly deflated a few months later when a report emerged from the Office of Scientific Integrity to the effect that Imanishi-Kari had indeed committed fraud, both at the time of *Weaver et al.* and later, when producing fabricated evidence to back up her initial claims of innocence. Further, the past she had constructed for herself was beginning to unravel; people who'd worked with her in earlier years began to come forward with their own doubts, and it seemed she had doctored her *curriculum vitae*, adding at least one degree that she did not possess. As it was Baltimore who had turned the affair into a *cause célèbre*, it was he who took the brunt of the scientific community's anger.

And yet somehow his clarion call against "meddling politicians" lingered on. As we'll see later, there's every reason to be nervous of politicians meddling with science, imposing upon it their own ideologies; but this is beside the point. The function of Dingell's Subcommittee was to ensure, with extensive scientific advice, that taxpayer money was not being squandered on foolhardy * or fraudulent research: it is the entitlement of any customer to resist being swindled. It is also at least arguable that, if science manifestly fails in its claimed ability to police itself, then other police must be called in.

Even more alarming than that the false call to arms lingered is that so did the coverup. Imanishi-Kari took her case to the Departmental Appeals Board of the Department of Health and Human Services, who eventually, in 1996, found that, despite presenting complete forensic evidence, the Office of Scientific Integrity had not *proved* her guilty of misconduct. Thanks to this legalistic nicety she was once more able to claim federal grant money and was reinstated in her position at Tufts; it was and is widely assumed that her name had been cleared. Baltimore, who in 1991 had been pressured into standing down from his recently acquired position at Rockefeller University, is currently the Robert A. Millikan Professor of Biology at the California Institute of Technology and received the National Medal of Science in 1999. He is still rightly regarded as a great scientist – he won the Nobel Prize for Physiology or Medicine in 1975 – and he has been a courageous and outspoken opponent of the Bush Administration's many genuine corruptions of US science. That his reputation will be forever tarnished by his obduracy (which can be read as an admirable determination to stand by a junior colleague) in the Imanishi-Kari case neither takes away from his scientific

* Within which single word lies, of course, an entire lengthy debate!

achievement nor assists science: the next time a junior scientist makes a discovery of fraud, the honourable impulse to whistleblow will be severely tempered by recollection of the treatment Baltimore and the rest of the scientific establishment dished out to O'Toole.

It should be noted that in *The Baltimore Case: A Trial of Politics, Science, and Character* (1998) Daniel J. Kevles presents an exactly contrary view of events: Imanishi-Kari was the innocent victim, Baltimore a righteous defender, O'Toole a fantasist, Maplethorpe a slimy *agent provocateur*, and Stewart and Feder witchhunters.

The Imanishi-Kari case was not Baltimore's first brush with fraudulent scientists, although there's no question that he played any part in the frauds of Mark Spector and John Long: he merely collaborated on research with them for relatively short periods of time prior to their exposure. The Spector affair we've already noted. John Long was from 1970 a resident at Massachusetts General Hospital, training under Paul C. Zamecnik (b1912). Zamecnik was researching Hodgkin's disease, and thus trying to culture Hodgkin's disease tumours. Most of the cell lines were fairly short-lived, but Long managed to establish several permanent ones, a feat that brought him much renown – including the collaboration with Baltimore – as well as promotion and grant money. His problems began in 1978 when a research assistant, Steven Quay, gained an anomalous result while working with one of Long's cell lines. Long repeated the experiment in lightning-quick time, and got a more expected result. Baffled by the whole affair, Quay shrugged it off. A year later, however, his suspicions still clung, and in due course it was proven that Long had faked his repeat of the experiment.

Long resigned from Massachusetts General Hospital, and that was when the *real* trouble started. Without Long there to perform the experiments on his cell lines, other experimenters found that they, too, were getting odd results. Further analysis showed one of the cell lines came from a patient who had not been suffering from Hodgkin's disease while the other three were not human at all, but came from an owl monkey! It is not uncommon for cell cultures to be contaminated by cells from elsewhere with the latter taking over the culture; considerable precautions are normally taken to prevent just this. No one knows how Long's cell lines became contaminated; but it seems impossible he was not at least aware of something being wrong. Otherwise, why would he have faked his repeat of Quay's experiment, not to mention various other bits of chicanery that turned up when his peers began to subject his published papers – shamefully, for the first time – to detailed scrutiny?

COLD FUSION

March 23 1989 brought news that stunned the world. The international scientific community was thrown into a frenzy; more publicly, so was the international financial community, while the news media subjected everyone on the planet to a sustained blitz. Two scientists working at the University of Utah, Stanley Pons and Martin Fleischmann, announced they had discovered a technique for cold fusion.

All other things being equal, a nuclear fusion reaction that doesn't require vast establishments and colossal investment need not be very efficient for it to make commercial sense. Hydroelectricity, for example, is not efficient at all yet makes plenty of sense because, once the equipment is in place, the running costs are minimal and the fuel costs zero. Similarly, to show the viability of cold fusion as a technology we don't have to show that it works *well*, just that it works *at all*. Little wonder, then, that there was such a furore in the wake of that announcement by Pons and Fleischmann: not only did they claim to have cracked the nut of cold fusion, they maintained they had done so at room temperature using equipment most of which might be found in the average school chemistry lab. The amounts of energy they were talking about were small, but then so was the scale of the experiment.

Physicists and chemists worldwide dropped everything

to try to replicate the experiment. Some were impatient with this confirmation process, and such hotheads included many financial speculators and a truly shameful number of politicians. The Utah State Legislature promptly hurled four and a half million dollars at Pons and Fleischmann. The US Office of Naval Research chucked in an initial $400,000. It looked as if at least tens of millions of dollars might be on the way from the US Government. When initial reports from other scientific researchers seemed favourable towards the Pons/Fleischmann results, further financing from industry seemed a foregone conclusion.

But trouble was brewing for the two chemists and their most ardent supporter, the University of Utah. Pons and Fleischmann had chosen to avoid the normal painstaking process of peer-reviewed publication and instead to use newspapers and TV as their mouthpiece. Whatever spin the various interested parties tried to put on this decision, it was to say the least suspicious.* Furthermore, while some early efforts to replicate the results seemed to hint at confirmation, a disturbing number of others did exactly the opposite – and soon the negative reports became the majority. Matters were not helped by some apparent belated fudging of figures on the part of a desperate Pons and Fleischmann. There were also astonishingly crass attempts by the University of Utah (who promptly disclaimed responsibility when this matter became public) to use the weapon of threatened litigation to silence critics – more than anything else this destroyed Pons's and Fleischmann's credibility. With the onslaught of adverse experimental evidence came demolitions of the theoretical underpinning of Pons's and Fleischmann's research.

One aspect that was downplayed by Pons and Fleischmann and their supporters was that deuterium fusion is not quite the "clean" energy source the public imagines: their putative cold-fusion cell would in fact be highly radioactive. Later, under pressure, they would try to explain away this paradox; at the time, the joke being surreptitiously repeated in the Physics Department at the

---

* It is still a complete mystery as to what on earth possessed the two men to act in this way. It's accepted they were at the time fully confident of their results, so it seems crazy they didn't follow the orthodox procedure.

University of Utah was: "Have you heard the bad news about Pons's lab assistant? He's perfectly healthy."

We shouldn't forget, however, that at least a few researchers believed they'd been able to duplicate Pons's and Fleischmann's results – and, in the years since the main furore died down, others have joined their ranks. The two Utah professors likely did not, as they thought, discover cold fusion; but they may have discovered *something* – and this something, whatever it might be, has not yet been fully investigated.

Similar doubts surround the claims made in 2002 by Rusi Taleyarkhan, then of the US Department of Energy's Oak Ridge National Laboratory, Tennessee, and later of Purdue University, Indiana, to have attained cold fusion. His team bombarded a beaker of chemically altered acetone with neutrons and then with sound waves to create bubbles; when the bubbles burst, the team reported in *Science*, fusion energy could be detected. Other teams, however, have had difficulty replicating the results – including Taleyarkhan himself. Working at Purdue, he finally announced in 2004 that he had done so using the uranium salt uranyl nitrate. Brian Naranjo of the University of California, Los Angeles, who in 2005 reported that his team there had attained cold fusion by heating a lithium crystal soaked in deuterium gas, analysed Taleyarkhan's results and concluded that the Purdue scientist had detected not cold fusion energy but leakage from some other radiation source present in the laboratory. If so, it seems a rather elementary mistake for Taleyarkhan to have made. More gravely, some of Taleyarkhan's Purdue colleagues began raising complaints about various aspects of the experiment. Purdue University initiated an investigation, but in February 2007 reported there had been no misconduct by Taleyarkhan.

THE SCHÖN CASE

The 21st century, though as yet only a few years old, has already witnessed several spectacular exposés of scientific fraud.

Dr Jan Hendrick Schön (b1970) was working at the Bell Laboratories in New Jersey, US, when he came to prominence in 2001. In a paper published in *Nature* in 2001,

"Self-Assembled   Monolayer   Organic   Field-Effect Transistors", he announced (along with junior co-authors H. Meng and Zhenan Bao) that he had managed to create a transistor on the molecular scale. This was an astonishing breakthrough: it was not just that electronic components could be made smaller but that organic-based chips could be made on a scale so small that silicon itself breaks down. The promise was of electronics that were smaller, hugely cheaper, and   far faster. It was headline news, and Bell Labs basked in   the glory.

Schön received the Otto-Klung-Weberbank Prize for Physics and the Braunschweig Prize in 2001, and the Outstanding Young Investigator Award of the Materials Research Society in 2002, but in fact almost immediately after the publication of that "breakthrough" paper people began raising red flags: not only did some of his data seem too good to be true, some of it seemed to have been duplicated from one experiment to another. In May 2002 a heartily embarrassed Bell Labs set up a committee to examine Schön's work, and they found considerable evidence of manufactured data and the like. (They also found his many junior colleagues blameless.) Bell Labs fired Schön in September 2002 when the committee delivered its report, while the editors of the more responsible scientific journals began the task of re-examining the papers by him that they had innocently published; by March 2003 *Nature and Science* had between them retracted no fewer than 15 of his papers because of suspect or bogus content.

THE CLONING OF A HUMAN BEING

But Schön's case in retrospect seems a mere warm-up for the next great science-fraud scandal of the 21st century, the cloning scandal centred on the South Korean biomedical scientist Hwang Woo Suk (b1953). In his native land, Hwang was not just a prominent scientist but a *star*. His field was cloning, and in February 1999 he announced he had created the world's fifth cloned cow; curiously, he failed to publish a scientific paper on this experiment. The same pattern of nonpublication was repeated a few months later when he announced he had cloned another cow. For the next few years he regularly claimed genetic-engineering breakthroughs, including the creation of a cow resistant to BSE. In 2004 his latest announcement concerned not food animals but humans: he had cloned a human cell, creating embryonic cells by inserting an adult cell's nucleus into a human egg. The following year, through a paper published in *Science*, he told the world he and his team had succeeded in creating 11 stem-cell lines from different individuals. He had improved the process of creating stem cells so radically that it seemed the great goal of obtaining stem cells that exactly match the individual's would soon be attained; this is important because, if a patient could be treated using his or her own cloned tissues, the risk of tissue rejection would (presumably) be entirely obviated. Whole cascades of cures seemed to come into view.

Soon the doubts began. Some young South Korean scientists had been unpicking Hwang's published work and posting the mistakes they found on the internet; mainstream print and TV journalism in South Korea had likewise become sceptical. However, the Western scientific establishment, the Western media and most importantly of all the South Korean Government took far longer to catch on. The crunch came when the young Korean scientists published to the internet two identical photographs taken from two separate papers written by Hwang, each purporting to show a different type of cell. Although Hwang was swift to claim the duplication was a mere accident – the wrong photo had been sent with one of the papers – his critics regarded it as evidence of deliberate fraud, and suggested he'd never cloned an adult human cell at all, instead deriving his stem-cell lines in the usual way, using spare embryos from fertility clinics. There were questions, too, about one of the illustrations to Hwang's 2004 paper, in which it seemed a supposed recording trace had been at least augmented by hand, if not completely an artifice. Soon individual members of Hwang's staff were going public about his methods and confessing that to their knowledge some at least of his 11 claimed stem-cell lines did not exist.

On November 24 2005 Hwang held a press conference to admit to various ethical lapses in his research: he had been blinded, he said, by his zeal to aid humankind. Despite its confessional tone, the press conference had what was presumably its desired effect: the drumming up of huge popular support for Hwang among the South Korean population and, significantly, among the nation's lawmakers. The weekly TV program *PD Su-Cheop*, which two days earlier had broadcast a (factually accurate) segment severely critical of Hwang's work, was pulled off the air. (Its cancellation was reversed some weeks later.) On December 17 2005 Seoul National University launched an urgent probe into Hwang's researches, seeking evidence of fraud; just 12 days later, on December 29, the university announced that all 11 stem-cell lines claimed by Hwang had been fabricated. The following January 10 the university announced that the data in Hwang's two *Science* papers of 2004 and 2005 were likewise fabricated. On January 12 Hwang held another press conference, saying the results had indeed been falsified but blaming others on his team, who he said had deceived him. By that time few were interested in his excuses. On March 20 he was fired from the Seoul National University, and on May 12 he was charged with fraud and embezzlement. (The trial continues at the time of writing.)

The case led to the formation in early 2006 of the Hinxton Group, set up to try to ensure that similar cloning frauds did not occur in the future. The scientific community's self-critical post mortem on how Hwang had been allowed to get away with so much for so long was extensive; in early 2007, for example, Donald Kennedy, Editor-in-Chief of *Science*, announced that in future, for "high risk" papers, the journal would no longer be content with the peer-review process alone but would subject such papers to additional and intensive editorial scrutiny before publication. A further spur had come with the 2005 *Lancet* publication of a faked paper by Norwegian physician Jon Sudbo; too late was it noticed that over 25% of his invented "patients" had the same birthdate. In 2006 the Research Integrity Office was set up to counter such malpractice in biomedical work.

The ripples of the Hwang case continue to expand. One of Hwang's team, Jong Hyuk Park, by now working at the University of Pittsburgh, was in 2007 barred for three years from receiving US federal funds after it was discovered that he'd faked two figures in a paper that he and colleagues had been planning to submit to *Nature*.

The field of human cloning has attracted several other dubious claimants.

Raëlianism is a prominent UFO religion, founded in 1973 by sports journalist Claude Vorilhon (b1946) and now with tens of thousands of adherents worldwide. Raëlians believe very strongly in the merits of human cloning, especially as a way in which infertile and gay couples may have children that are genuinely their own. To this end, in 1997 the Raëlian Movement founded the company Valian Venture Ltd, which conducts research on human cloning and, through its project called Clonaid, a service to help couples who wish to have cloned offspring and have the necessary $200,000 fee.*

In December 2002 Clonaid's CEO Brigitte Boisselier claimed the company had cloned a human child, called Eve; in January 2003 Boisselier claimed a second cloned child had been born and that several others would soon follow in different parts of the world. These announcements naturally created international headlines, but further details (and evidence such as DNA samples) have not been forthcoming, the excuse being offered that Boisselier might find herself in jail for having violated various countries' anti-human-cloning laws; it's therefore generally assumed the claims were false – or perhaps a "metaphysical truth".

Hwang's and Boisselier's claims were presaged more than two decades earlier by David Rorvik's book *In His Image: The Cloning of a Man* (1978), which described how some of the world's top geneticists had been recruited by him on behalf of a billionaire (identified only as "Max") to clone a male heir for the latter, and had succeeded in so doing. Since Rorvik (b1944) was no wild-eyed conspiracy theorist but a science journalist with a respectable track record, this story was taken seriously by the press and public, even though specialists were less believing. The claim began to unravel when the Oxford University developmental biologist J. Derek Bromhall – who served as technical consultant on the movie *The Boys from Brazil*

---

* The focus of the Raëlian Movement on cloning is not merely an opportunistic one: it is a religious duty, because according to their doctrines it is only through cloning that humans can attain eternal life and thus become one with the Elohim.

(1978) – recognized his own doctoral thesis as the basis for the cloning details described in the book and took author and publisher to court: Rorvik was unable to offer any evidence in support of his account, and so the defence case collapsed. It's thought Rorvik's motive in writing the book may have been, recognizing that human cloning could become a reality in the not-too-distant future, to kick-start the necessary public debate about the ethics of it before it happened rather than after. Even so, Rorvik still continues – as recently as 1997 in an article in *Omni* – to insist his story was genuine.

PILTDOWN MAN

Probably the greatest palaeontological embarrassment of all concerned a fossil assemblage that appeared to be the remains of a Neanderthal-style hominid. The assemblage was "discovered" in early 1912 in a gravel near Piltdown Common, Fletching, Sussex, UK, and was christened Piltdown Man, or *Eanthropus dawsonii*.

The historical stage had been set for the "discovery". British palaeontology seemed to have been in the doldrums for decades. While workers on the Continent were unearthing exciting fossil hominids, their UK counterparts were lagging sadly behind. Patriots were desperate to convince themselves that, even in the field of prehistoric hominids, British was best.

The solidity of this mindset among establishment scientists can be assessed by consideration of the reception given to the discovery of "Galley Hill Man". In 1888, at Galley Hill in Kent, a very modern-looking human skull was discovered by chalk workers and excavated by two local amateurs, Robert Elliott and Matthew Heys. After their initial examination of it, the skull found its way to Sir Arthur Keith (1866–1955), head of the Royal College of Surgeons. Basing his assessment on the stratum in which the skull had been found, Keith was swift to go out on a limb and say it was of Neanderthal vintage, if not older. This seemed to confirm to him the presapiens hypothesis of Pierre Marcellin Boule (1861–1942) and others that humans indistinguishable from moderns had existed in eras of enormous antiquity. In fact the skeleton proved to be that of a Bronze

Age individual shoved down unwittingly into lower strata by later chalk digging. Those who had earlier shyly suggested the fossil skeleton might be comparatively recent had been, at least at first, generally dismissed – if not ridiculed – as "overcautious".

So, when amateur geologist Charles Dawson (1864–1916) made his "find" near Piltdown Common, he was assured of at least a tolerant reception. His further searches were assisted by some quite eminent figures of the day, notably the distinguished geologist Sir Arthur Smith Woodward (1864–1944), who examined the skull and proclaimed it to be that of a hominid, and the French geologist, palaeontologist, philosopher and Jesuit priest Pierre Teilhard de Chardin (1881–1955). Their haul included a fossil hominid jaw and various bits of animals otherwise unknown to the English south coast.

Despite occasional sceptical comment, Piltdown Man was for over 40 years accepted as the UK's great palaeontological discovery, although it was difficult to fit it into the accepted "map" of hominid evolution – indeed, that was the very reason Piltdown Man was so interesting.* Not until 1953 was the "fossil" shown – by J.S. Weiner, K.P. Oakley and W.E. Le Gros Clark – to be part of the skull of a relatively recent human being plus the jaw of a 500-years-dead orangutan, the remains having been stained to give the illusion of antiquity and to make them better conform visually with each other. The success of the deception owed less to the rudimentary nature of scientific dating techniques in the early decades of the 20th century, more to the fact that no one really wanted to believe other than that the Piltdown Man was genuine.

The identity of the hoaxer is still not known for certain. In *The Piltdown Men* (1972) Ronald Millar points the finger at Sir Grafton Eliot Smith (1871–1937), the distinguished Australian anatomist and ethnologist, who was in the UK at the time and who was well known for his sense of fun. In a famous essay, "The Piltdown Conspiracy" (1980), Stephen Jay Gould turns the spotlight on the role of Teilhard de

* At a less scholarly level, the US clairvoyant Edgar Cayce (1877-1945) was given the psychic insight that *Eanthropus dawsonii* was in fact an Atlantean, one of a number of refugees from the doomed continent who had made their home in the UK.

Chardin. However, Charles Dawson – who on occasion had actually been seen experimenting on the staining of bones – has generally been regarded as the prime suspect.

There is, though, yet a further wrinkle to the story. In the late 1970s a box was discovered in the British Museum's attic containing bones stained in the same way as Piltdown Man's had been. The box had belonged to Martin A.C. Hinton, later the museum's Keeper of Zoology but at the time a volunteer worker there. In 1910 he had been denied a research grant application by Smith Woodward; just two years later the Piltdown remains were unearthed. Revenge? Hinton is today often cited as a likely candidate. However, this may be a misjudgment. A more probable hypothesis, backed by anecdotal evidence, is that Hinton was initially just an observer of the hoax, and suspected Dawson as the perpetrator. As he watched Smith Woodward's claims for Piltdown Man escalate, though, Hinton decided to prick the bubble while at the same time making Smith Woodward look a fool; he therefore planted some extra remains that were all too obviously bogus: a reported example was a cricket bat hewn from an elephant's leg-bone. To Hinton's horror, however, when the cricket bat was unearthed Smith Woodward proudly announced it to the world as a prehistoric artefact of hitherto unknown type. At that point Hinton threw his hands in the air and gave up.

A decade after the Hinton discovery, yet another suspect emerged: William Sollas (1849–1936). He was fingered in a posthumously revealed tape recording made ten years earlier by J.A. Douglas (d1978). Both men had been professors of geology at Oxford University, Sollas at the time of the Piltdown Man sensation and Douglas succeeding him in the chair. Douglas could recall instances of Sollas purloining bones, teeth and similar items from the departmental stores for no specified purpose, and it's a matter of public record that Sollas thought Smith Woodward an overweening buffoon, such as it might be a pleasure to expose through a prank. Douglas's case is not strengthened by the fact that Sollas was far from alone in this opinion of Smith Woodward, and at this late stage it's impossible for us to evaluate the importance of Douglas's circumstantial evidence.

A further possibly relevant fact is that we now know Dawson had a secret career as a plagiarist. At least half of his critically slammed but popularly well received two-volume *History of Hastings Castle* (1910) was discovered in 1952 to have been copied from an unpublished manuscript by one William Herbert, who had been in charge of some excavations done at the castle in 1824. And nearly half of a paper Dawson published in 1903 in *Sussex Archaeological Collections* is reputed to have been lifted verbatim from an earlier piece by a writer called Topley, although in this instance the details are vague. But how could any evidence defame him when he had the supportive testimony of one Margaret Morse Boycott?

> Mr Dawson and I were members of the Piltdown Golf Club. Let me tell you this. He was an insignificant little fellow who wore spectacles and a bowler hat. Certainly not the sort who would put over a fast one.

FAKERY OF THE ANCIENTS

Palaeontological fraud on a large scale was unearthed in 2004, the figure at the centre of the case being Professor Reiner Protsch von Zieten (b1939) of the University of Frankfurt. He had built up an impressive record of sensational discoveries, notably his dating of a hominid skull found near Hamburg to 36,000 years old: so-called Hahnhöfersand Man was thus the oldest known German and a possible missing link between the Neanderthal line and that of modern humans, indicating not only that they co-existed but that they actually interbred. Other ancient human remains were of Binshof–Speyer Woman, which Protsch dated to 21,300BP, and of Paderborn–Sande Man, dated to 27,400BP.

A flamboyant figure, Protsch seemed also to be an archaeologist with a knack. However, Thomas Terberger of the University of Greifswald became suspicious, and sent the Hahnhöfersand Man skull to the radiocarbon dating unit at Oxford University for a second opinion. The verdict was that the skull was a mere 7500 years old – still interesting but hardly sensational. Dating of the other two human fossils revealed that Binshof-Speyer Woman was about 3300 years old, not 21,300, and, most embarrassing of all, that Paderborn–Sande Man had died *c*1750.

Terberger, in conjunction with UK archaeologist Martin Street, published a paper in 2004 reporting the awful truth, and the University of Frankfurt promptly began an investigation. More horrors soon turned up. The German police were especially interested to discover that Protsch had attempted to sell the university's entire collection of 278 chimpanzee skulls to a US dealer for $70,000. Under Protsch's orders, thousands of documents from the university's anthropology archives that detailed the vile human experiments of the Nazis in the 1930s had been shredded. There was nothing wrong with the dating of a "half-ape" skull called *Adapis*, already known to be of a species extant some 50 million years ago; however, Protsch reported that it had been found in Switzerland, far from its known territory, which made the finding significant. Alas, it now proved actually to have been dug up in France, like all the other *Adapis* fossils. The newspaper *Der Spiegel* discovered that the professor, who claimed to be the descendant of a Prussian nobleman – hence the "von Zieten" part of his name – was in fact simply Reiner Protsch, son of a Nazi MP, Wilhelm Protsch, which might go some way toward explaining the illicit document-shredding.

Unsurprisingly, in February 2005 the University of Frankfurt forced Protsch into retirement. An initial reaction from the palaeanthropological world was that large chunks of the accepted history of our ancestors might have to be rewritten, but later, when the dust had settled, it was generally realized that the damage done by Protsch's fraudulent dating was in fact only minor: his "finds" had been sensational but not terribly important.

A rather similar scandal erupted in Japan a few years earlier, this time concerning the discoveries of Shinichi Fujimura (b*c*1950), an amateur archaeologist and volunteer at scores of digs around Japan. Although he began his career in 1972, it was in 1981 that he made his first sensational "discovery": an artefact which he claimed to have found in strata that were up to 40,000 years old. After that there was no stopping him. Among professional archaeologists he became known as "God's Hands" for the magical knack he had of turning up significant items, and despite his lack of paper qualifications he was made Deputy Director of the Tohoku Palaeolithic Institute. It seemed that at every site where he worked he would come up with some significant artefact or another, and always the limits of Japan's known archaeological history were being pushed back farther and farther. This was important to many Japanese because it seemed the previously accepted hypothesis of the Japanese coming from the same stock as the Koreans would have to be discarded: the Japanese were of separate ethnic stock, just as the nationalists in Japan had always, in the teeth of existing evidence, claimed. Not only in Japan were various textbooks rewritten in the light of Fujimura's startling finds.

On October 23 2000 Fujimura held a press conference to announce that his team's latest discoveries were a whopping 570,000 years old. However, a couple of weeks later, on November 5, the newspaper *Mainichi Shimbun* published some photographs of Fujimura digging a hole at the site and planting artefacts in it; this photo, the newspaper said, had been taken the day before the sensational find. That same day Fujimura held another press conference, this time to confess to gross archaeological fraud. Many of the artefacts he had unearthed over the years had come, he said, from his own personal collection and been planted by him. He was careful to distinguish these from others he insisted were genuine; later researches have, however, indicated that these too are almost certainly fraudulent. Enormous shockwaves ran through Japanese archaeology, and one distinguished archaeologist, Mitsuo Kagawa (*c*1923–2001), was so traumatized by accusations he might have been party to the fraud that he committed suicide. Meanwhile, of course, the textbooks had to be rewritten once more to reflect the pre-existing hypothesis.

Fujimura, in his turn, was not without his precursors. Most notable among these was perhaps Viswat Jit Gupta, a professor at the Punjab University of Chandighar whose fraudulence was exposed in 1989 in *Nature* by the Australian palaeontologist John Talent. For some quarter-century the palaeontology and stratigraphy of the Himalayas had been, to say the least, puzzling: fossils were turning up that seemed to have no business there, and they were turning up in inexplicable strata – old fossils in higher strata than more recent ones, with no apparent geological disruption to explain how this could have come about. Talent pointed out that all the anomalous finds had been made by one man, Gupta, who had detailed them in over 300 papers. Quite how many further papers were corrupted through having built on Gupta's "work" is a number hard to contemplate. It seems he merely appropriated fossils whose disappearance would go unnoticed, planted them anywhere between Kashmir and Bhutan, and then, well, dug 'em up again.

ARCHAEORAPTOR

The November 1999 issue of the magazine *National Geographic* caused a sensation when it announced the discovery of a fossil that represented the missing link between dinosaurs and birds. In this article Christopher Sloan, the magazine's art editor, proposed the name *Archaeoraptor liaoningensis* for the fossil, whose stated history was that it had been discovered in China and smuggled out of that country to the USA, where it was bought at a gem fair in Tucson, Arizona, by Stephen Czerkas, who owns the Dinosaur Museum in Blanding, Utah. What Sloan did not mention in his article was that his own scientific paper on the fossil had been rejected by the scientific journals *Nature* and *Science*.

Czerkas did the decent thing and sent the fossil back to China, where it was examined at the Institute of Vertebrate Palaeontology and Palaeoanthropology, Beijing. A researcher there, Xu Xing, soon spotted a strong resemblance between the rear half of the supposed fossil and that of an as yet unnamed dinosaur which he had studied; the front half, however, was quite different. Eventually the fraud was traced to a Chinese farmer, who had created the chimera by fixing bits of two different fossils together with very strong glue. The dinosaur whose rear Xu had recognized is now called *Microraptor zhaoianus*; the front part of "*Archaeoraptor*" was identified in 2002 as belonging to the ancestral bird species *Yanornis martini*.

Earlier, in the 1980s, there had been suggestions that the science of palaeontology as a whole might have been fooled by what would rank as perhaps the most successful

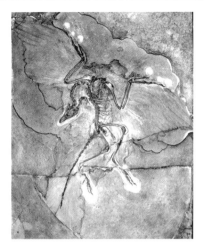

fraud of all time. *Archaeopteryx*, the recognized interim between reptiles and birds, is known from a scattering of fossils found in Bavaria; these are believed to date from about 150 million years ago. Only a few isolated bone fragments have been discovered from the next 50 million years to link *Archaeopteryx* to the main sequence of bird evolution found in fossils dating from about 100 million years ago onwards – in other words, there's a good line of fossils, showing clear development over time, but it starts much later than the date registered by *Archaeopteryx*. The orthodox explanation for this dearth of early fossils is that bird fossils are rare overall: bird bones are small and fragile, and can easily be overlooked. Moreover, because of the way fossils are formed, aquatic birds are much more likely to be discovered than land-based ones.

To some people – notably Sir Fred Hoyle (1915–2001) – this all sounded a little suspect. If the conventional explanations were true, then surely we should count ourselves improbably lucky to have discovered even a single specimen of *Archaeopteryx*, the oldest of all in the bird lineage. Yet we have 10 – and all 10 come from the same region of Bavaria. Besides, the fossils are exceptionally good ones: in places you can make out not just the skeleton but the very feathers!

The first *Archaeopteryx* fossil, found in 1860, was of a

single feather; it was followed just a few months later by a complete specimen, purportedly given to Bavarian physician and amateur fossil-collector Karl Häberlein in 1861 as payment for medical services, and sold by him to the British Museum for the then princely sum of £700. It is today regarded as the most valuable fossil in the world, with only a replica being put on display and the precise location of the original kept a closely guarded secret. Some 16 years later, in 1877, Karl's son Ernst produced another, arguably even better specimen; he got yet more for it, the industrialist Ernst Werner von Siemens (1816–1892) paying him £1000 and in due course donating the fossil to the Humboldt Museum, Berlin. These are the only two *Archaeopteryx* fossils where the evidence of feathers is unequivocal. Following a close examination of the British Museum specimen, physicist Lee M. Spetner, one of Hoyle's crew of sceptics, concluded that the impressions of the feathers were a later addition: these impressions seem to be in a far finer-grained rock than is the rest of the fossil, and Spetner speculated that a cement had been made using powdered local limestone, so as to match the original rock, with impressions being made in this cement using nothing more arcane than chicken feathers. Photographic evidence lends support to his hypothesis, but it is still regarded as controversial. A further matter of note is that the slope of the hip bones in the other *Archaeopteryx* fossils is typically reptilian, whereas in the two sold by the Häberleins it is more typically avian.

Could it be that the crafty yeomen of Bavaria were concocting what the palaeontologists and museums so obviously wanted and selling it to them via respectable frontmen? There was quite an industry in faked fossils and archaeological items in that part of Europe during the 18th and 19th centuries, but surely a fraud would have been exposed long ago. Wouldn't it?

THE CHEN JIN CASE

The People's Republic of China is notorious for the *laissez faire* attitude it adopts toward technological piracy and intellectual copyright, right down to the open bootlegging of

DVDs. However, in recent years there has been a significant push by the country's government to encourage R&D at the cutting edge of technology, especially in computer and information science, so that China may establish itself as a superpower in the fullest sense of the word. This, coupled with the relaxation of state communism so that innovators can benefit substantially from their innovations, has put hitherto-unknown pressures on scientists to succeed: glory and riches are the carrot, while the threat of state disapproval is the stick. Observers within China reckon that the corruption of science has in consequence become endemic there.

A particularly high-profile case emerged in May 2006 with the downfall of Chen Jin (b1968), one of China's most acclaimed computer scientists. Chen had studied at Shanghai's Tongji University and then gained his PhD at the University of Texas at Austin. He worked for a while at Motorola's research centre in Austin, then returned to China to work for the same company's division near Shanghai. He next accepted a post at Jiaotong University, where in 2003 he created, he claimed, the Hanxin, a chip capable of processing 200 million instructions per second. Jiaotong University promptly established its own microelectronics school, with Chen as its head. Eager to eliminate the need for China to import all its chips – an expenditure reputedly running into billions of US dollars – the Ministry for Science & Technology, the Ministry for Education and the National Reform & Development Commission financed Chen heavily. He was able, too, to set up a string of companies designed to exploit his achievements for his own profit. In 2004 the Hanxin II and the Hanxin III followed, both capable of even faster speeds.

Toward the end of 2005, however, disgruntled former colleagues of Chen's began posting to the internet, then writing to the government and to the authorities at Jiaotong University, reporting that much or all of Chen's research had been faked: the supposed Hanxin chips were in fact Motorola's chips, carefully doctored to remove the Motorola identification. Once Chen had determined the specifications of the Motorola chip, it was a simple matter to pass them on to manufacturers as if they were his own invention.

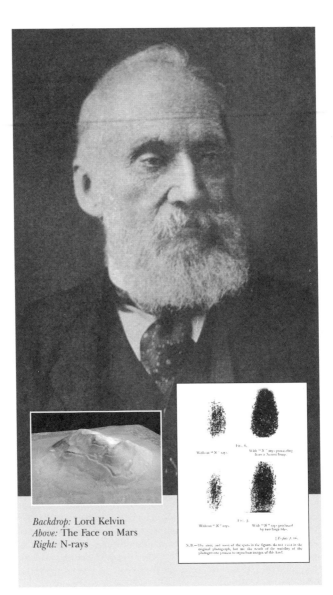

Backdrop: Lord Kelvin
Above: The Face on Mars
Right: N-rays

CHAPTER 2

SEEING WHAT THEY WANTED TO SEE

———————— ⋖◉⋗ ————————

Lord Kelvin (1824–1907) was one of the greatest physicists of all time, but he also had a habit of being comprehensively wrong about things. Two of his ringing predictions have achieved classic status: "Heavier-than-air flying machines are impossible" and "Radio has no future." He proved evolution was bunk through calculating that the sun wasn't old enough to allow for the long time periods required by Darwinian evolution. Of course, Kelvin could have no clue that the sun "burns" through nuclear fusion, not by mere combustion; where he erred was in rejecting Darwinian evolution purely because he found the notion philosophically intolerable and then grabbing the first proof he could find that his revulsion represented science.

Almost as celebrated at the time was his accusation that Wilhelm Roentgen (1845–1923), who in 1895 announced the discovery of X-rays, was guilty of scientific fraud. Kelvin's argument was that the cathode-ray tube, the source of the rays, had been in use by physicists for a full decade before the purported discovery. If X-rays existed, someone would already have noticed them. Well, someone just *had* . . .

Irving Langmuir (1881–1957), winner of the 1932 Nobel Chemistry Prize, in 1934 coined the term "pathological science" in the context of the researches J.B. Rhine (1895–1980) was carrying out on ESP. Langmuir's informal definition of the term was "the science of things that aren't so" – in other words, the application of science to a phenomenon that doesn't exist, or in pursuit of a hypothesis that is nonsensical. The pathology lies in the self-deception: the researchers or theorists are incapable of standing back and observing that they're engaged in a fool's errand. In this chapter our focus will be on such science, although always with the thought in the back of our mind that sometimes it isn't obvious at the time just what science is pathological and what is merely new and unexpected.

WOO-WOO SCIENCE

John Taylor (b1931) is a distinguished mathematician. He is not, however, a distinguished conjurer. When in November 1973 he was asked to take part in a BBC television programme with Uri Geller (b1946), who was making his first big international splash as a psychic spoon-bender, Taylor approached the encounter with the sceptical attitude of a scientist and left it as a gull, dazzled as much as any ordinary punter might be by what he had seen Geller do. When the BBC's phone lines were clogged by callers reporting how, as a result of watching Geller perform, they'd discovered paranormal powers of their own, Taylor reckoned that here was a phenomenon which, despite all the biases against it of science, merited proper scientific investigation. He set up experiments in which children apparently bent cutlery that was sealed in glass tubes that were firmly corked at each end. It was all a most impressive demonstration of psychokinetic power.

The only trouble with Taylor's investigation was that it seems never to have occurred to him that experimental subjects might *cheat*. He therefore didn't think to incorporate any proper checking mechanisms; he even allowed some of the children to take home the tube-enclosed cutlery to give them the opportunity to focus their psychic energies on it outside the possibly alienating confines of the laboratory. When others decided to do some elementary checking, by the simple means of spying on the children, they were able to do such things as film some of the little darlings removing the corks from the ends of the tubes and bending the spoons against the side of the table or underfoot . . . This the kids were able to do even when the experimenters were in the room by employing the elementary conjuring stratagem of distracting the experimenters' attention during the crucial moments.

Eventually enough evidence of such chicanery emerged that Taylor himself recanted, publishing his conclusion –

that there was nothing in the field of the paranormal worth investigating – in the book *Science and the Supernatural* (1980). His primary error lay in not realizing from the outset that a scientist who strays into an unfamiliar scientific (or in this instance pseudoscientific) field is little better equipped to evaluate it than any of the rest of us: for example, a biologist's assessment of climate science is of little more value than yours or mine . . . and possibly of less, in that we don't bring to the subject as much potentially misleading baggage in the form of false confidence. In the case of the paranormal, the appropriate "science" is conjuring, not mathematics.

In a sense, Taylor was continuing in a long and distinguished tradition of scientists being fooled into giving charlatans credit through the misapplication of the Scientific Method. A weakness of the Scientific Method is that it assumes the integrity of all participants – and, if there were no exploitative crooks in the world, why should it not do so? – whereas the whole field of the paranormal is infested with participants who have no integrity, whose purpose is to deceive. James Randi (b1928) and, long before him, Harry Houdini (1874–1926) have been two to point out forcefully that scientists should, when investigating anything to do with the supernatural, have as their vital colleagues professional conjurers, whose training in fakery better enables them to spot it.

Spiritualism famously ensnared the physicists Sir William Crookes (1832–1919) and Oliver Lodge (1851–1940). Thomas Edison (1847–1931) was so taken in by the Theosophy of Helena Blavatsky (1831–1891) that he tried to invent a machine for communicating with the dead. The psychologists Gustav Fechner (1801–1887), Ernst Weber (1795–1878) and Hans Eysenck (1916–1997) fell into similar paranormal traps, while it's hard to know where to start with psychoanalysts like Carl Gustav Jung (1875–1961) and Wilhelm Reich (1897–1957). Among Nobel laureates we can mention the French physiologist Charles Robert Richet (1850–1935). At a more serious level of involvement we have of course J.B. Rhine, who devoted what might have been a sterling research career to ultimately fruitless investigations of ESP, and today Russell Targ and Harold Puthoff. And a 1997 survey found that 40% of US scientists believed they could communicate with God and that He was capable of answering their prayers.*

THE GREAT RABBIT SAGA

Even within their own disciplines scientists can make egregious errors, as happened to the UK physicians John Howard and Nathanael St Andre (1680–1776); Howard was merely a local doctor and midwife, but St Andre was surgeon-anatomist to George I. The case that brought both men low was that of a Godalming woman, Mary Toft (*c*1701–1763), who in 1726 started giving birth to rabbits . . . or so she claimed. Howard – who'd earlier been amazed on delivering her of parts of a pig! – was first to examine her, and believed the phenomenon genuine; it was as a result of his letters to other scientists that Toft rapidly became a celebrity and St Andre was sent to investigate.

According to Toft, the strange births were a consequence of her having craved rabbit meat during a recent pregnancy that had ended in miscarriage. She was brought to London, and drew big crowds . . . but also the attention of the prominent surgeon Sir Richard Manningham (1690–1759). Manningham announced that he wished to

* To be fair, the same survey showed that a whopping 87% of the US adult population believed the same thing, so a scientific education is obviously not *entirely* wasted.

operate on Toft to find out what was going on in her womb, and this seems to have been the final trigger that evoked a confession from her. Toft admitted her husband had bought the dead rabbits and she'd inserted the furry little corpses into herself, then "birthed" them at appropriate moments – her aim being to become such a celebrity that King George would grant her a pension. In fact, she became an involuntary guest of King George for a while, and the two physicians became a laughing stock. Jonathan Swift's *The Anatomist Dissected* (1727) was one of several satirical pamphlets on the subject, while the affair is also celebrated in William Hogarth's print *Cunicularii, or The Wise Men of Godliman in Consultation* (1726).

LETTERS FROM DISTINGUISHED PERSONS

The error of the distinguished French mathematician Michel Chasles (1793–1880), a member of the French Academy of Sciences and Professor of Geometry at the Imperial Polytechnic of Paris, was to stray not just into another scientific discipline but into the field of historical documents. In 1851 he met a plausible character called Denis Vrain-Lucas (1818–1882), who told him he had come across some letters of historical interest. Their writers were prominent figures from French history and literature – Molière, Rabelais, Racine, Joan of Arc and Charlemagne, to name just a few. Chasles eagerly bought the letters, and urged Lucas to contact him should any more of the kind turn up. Naturally, they did – letters from Julius Caesar, Mary Magdalene, William Shakespeare and countless others, all of whom, strangely, had put their thoughts to paper in French. It was even more remarkable that some of them had put their thoughts to paper at all, because they were writing at a time before paper was invented . . . let alone the particular watermarked paper most of the correspondents had chosen to use.

Vrain-Lucas had served his apprenticeship, as it were, in the lucrative field of false genealogy. Thwarted through lack of educational qualifications in his desire to make a career out of librarianship, he'd taken a job with a Paris genealogical company whose practice, when documents were lacking to confirm rich clients' relationships to famous historical figures, was to forge them. Vrain-Lucas proved quite good at this. When the head of the firm retired, he left his protégé his collection of (genuine) autographs and documents – a handy toolkit for the aspiring literary forger. And then Vrain-Lucas found Chasles.

It took nearly 20 years before, in 1870, Chasles finally admitted he'd been had and took Vrain-Lucas to court. Even then, the charge wasn't one of selling him faked documents; the case was brought because Vrain-Lucas had failed to deliver 3000 autographs Chasles had paid for. At the trial it emerged that the next manuscript Vrain-Lucas had been planning to sell him was Jesus's own copy of the Sermon on the Mount.

Yes, it too was in French.

N-RAYS

In 1903 the prominent French physicist René-Prosper Blondlot (1849–1930) discovered N-rays, produced naturally by various materials, including many of the metals, but also by the human nervous system – notably, when people talked, by the part of the brain that controls speech, Broca's area. (He called them N-rays in honour of his employer, the University of Nancy.) His findings were confirmed by other French scientists, although outside France experimenters had difficulty in reproducing his results.

Using adapted spectroscopic equipment – in which the lenses and prisms were made of aluminium – Blondlot could project N-ray spectra. It was exceedingly difficult to discern the lines of the projected spectra: only those with finely attuned eyesight could do so, and the average, untrained observer was likely to see, well, nothing. The US physicist Robert W. Wood (1868–1955) attended several of Blondlot's demonstrations and came to the conclusion that he must be an average, untrained observer – or perhaps not. Wood performed various tests and soon decided instead that N-Rays were a figment of the French scientist's imagination. At one point, while Blondlot was describing a projected N-ray spectrum, concentrating on the lines in front of him, Wood surreptitiously removed the aluminium prism from the N-ray "spectroscope". Blondlot continued his description, unperturbed.

In 1904 the French Academy of Sciences bestowed on Blondlot the prestigious Leconte Prize. In that same year, though, Wood published an article recounting his experiences, and non-French scientists promptly stopped looking for the rays. Within France, however, physicists not only kept looking for them but also in many cases finding them. For some decades thereafter it was a source of nationalistic pride that French scientists could detect and study "their" rays while their counterparts abroad neglected this important field entirely. In the end, of course, stubbornness could withstand reality no longer, and the French followed the example of their foreign colleagues.

Blondlot and his compatriots sincerely believed they could see the N-ray spectra, and that Blondlot had made an important discovery. Part of what made them see what they wanted to see was undoubtedly nationalism: physicists in other countries had been discovering exciting new rays, making French physics look shoddy by comparison. Another large part of it may simply have been ambience: the climate in French physics became one in which the existence of N-rays was accepted as a given, and so naturally any physicist worthy of his (rarely, in those days, her) salt was able to see the spectra and conduct successful experiments to investigate the rays' properties.

ASTRONOMY DOMINOES

There are plenty of examples of astronomers managing to fool themselves by seeing what they want to see rather than what's actually there – hardly surprising since, throughout much of astronomy's history, observation was a matter of peering for long periods at objects that, even through telescopes, were incredibly faint. Exactly the same sort of thing happened in the early days of microscopy: as they squinted through their imperfect lenses at images that lurched in and out of focus, some 17th-century microscopists were even

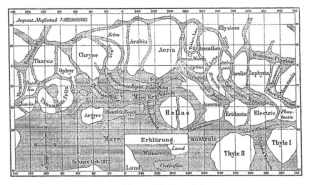

Schiaparelli's 'canals'

able to observe homunculi – tiny, fully formed human beings that supposedly inhabited what we would now call sperm cells.

The classic example of this phenomenon in astronomy is of course that of the "canals" of Mars. In light of observations he'd made during 1877–81, the Italian astronomer Giovanni Virginio Schiaparelli (1835–1910) declared he could see straight markings criss-crossing the planet's surface. He called these markings *"canali"*, meaning "channels"; the word was translated into English as "canals", with all the connotations the latter word brought with it. The assumption spread rapidly that Mars was inhabited by a widespread alien civilization. Not just canals were seen: there were also fluctuating markings that could surely be nothing else but seasonal variations in the Martian vegetation cover. It didn't take long before the two concepts were put together, so that the obvious purpose, aside from transport, for which the Martians built their canals – a huge endeavour, bearing in mind that no human enterprise at all would be visible to astronomers on Mars – was irrigation of their crops. Right up to the time that the first space probes began sending back photographs of the Martian surface it was not unrespectable to speculate about the nature of Martian vegetation. By that time, however, the notion of the canals had generally been abandoned.

The two chief champions of the canals hypothesis were Camille Flammarion (1842–1925), who believed that all worlds were inhabited and who had reported observing the growth of vegetation on the moon; and particularly the US astronomer Percival Lowell (1855–1916). For 15 years Lowell studied Mars, mapping the canals incessantly; his books *Mars and Its Canals* (1906) and *Mars as the Abode of Life* (1908) were enormously influential, not only among some astronomers but also, more significantly, among the general public. Both Nikola Tesla (1856–1943) and Guglielmo Marconi (1874–1937) claimed to have received radio signals from Mars. All of this was fantasy based on a mistranslation, on the limitations of earthbound telescopes, and on human psychology.

Less well known than the Martian "canals" fiasco is an embarrassment of the great astronomer Sir William Herschel (1738–1822). Herschel's strongest claim to fame is that he discovered (1781) the planet Uranus, the first "new" planet to be identified since Classical times. When he began studying the celestial object he'd identified he thought it was a comet approaching the sun; accordingly, his observational records show a slow increase in the object's size. In due course he realized the truth,  but still felt no need to reconsider his observations: clearly the planet was at a stage of its orbit where it was approaching the earth. Only once Uranus's orbit had been properly calculated did it became clear that, during Herschel's observational period, the planet was actually receding: an accurate record would have shown the mystery object getting, if anything, smaller.* There's no question of his deliberately faking his records. Herschel, like so many others before and since, simply saw what he expected to see.

Another great UK astronomer, Sir John Flamsteed

* In reality, since Herschel conducted his observations over a mere few weeks, it's improbable he could have detected any change in the object's size at all.

(1646–1719), had a theory whose proof would lie in Polaris being further from the celestial north pole in summer than in winter. He made the appropriate observations, and sure enough found this to be the case. Other astronomers repeated the observations and found no such effect (there is none). When confronted, Flamsteed blamed his instruments, which he thought must have introduced the error. It seems never to have occurred to him that he'd simply deceived himself.

Closer to our own time, in the 1910s and 1920s astronomers at the Mount Wilson Observatory, led by Adrian van Maanen (1884–1946), reported that the spiral nebulae could be seen to be slowly unraveling. At the time it was thought all nebulae were clouds of gas and dust within our own, unique galaxy. That some of them should display structure was a puzzle, resolved by the theory that the nebulae formed as ellipsoidal balls and then slowly disintegrated to become irregularly shaped clouds; van Maanen's discovery, based on close comparison of photographs made at different times, appeared to confirm this notion. Such was the prestige of Mount Wilson that none dared contradict the conclusion, and the astronomy establishment grew quite sharply defensive when questions were raised. Any dissent remained stifled until Edwin Hubble (1889–1953) showed in 1924 that certain nebulae – including those of ellipsoidal ("elliptical") and spiral form – are hugely distant galaxies; the data that van Maanen had recorded were quite simply impossible.

While there's no guilt in van Maanen and the others having deceived themselves when believing they could see incredibly small differences in the photographs, a strong finger of condemnation should be pointed at those in the astronomy establishment who bullied everyone else into going along with the false data even when these grew increasingly dubious. Among them was Harlow Shapley (1885–1972), who most certainly should have known better.

POLYWATER

In 1962 the Russian chemist N.N. Fedyakin, of whom little else is known, announced the discovery of what appeared to be a new form of water: it had a lower freezing temperature (about –40°C) than water's yet tended to solidify at a much higher temperature than this (about 30°C), boiled at about 250°C rather than the customary 100°C, was at least several times more viscous, and, depending upon circumstances, appeared to be about 40% denser or 20% less dense. This kind of water could be found when ordinary water vapour had condensed in very narrow glass or fused-quartz tubes; it was christened "anomalous water", "water II", "orthowater" and much later, by UK and US researchers, "polywater" – the last because it was assumed to be a hitherto-unknown polymer of water.

Much research into polywater was done by Fedyakin and colleagues at Moscow's Institute of Physical Chemistry. One of those colleagues, Boris Derjaguin or Deryagin (1902–1994), came to the UK in 1966 and lectured on polywater to the Faraday Society, prompting UK scientists such as J.D. Bernal (1901–1971) to repeat the experiments successfully. A couple of years later, the main activity of experimentation shifted to the US, where copious papers on polywater and its deduced properties were published until the early 1970s, when it was discovered, as spectroscopy became more sophisticated, that polywater was merely water contaminated by impurities from the tube walls.

In effect, there were two forms of polywater, depending upon whether the capillary tube used was of quartz or of glass. In the former case, tiny pieces of quartz from the tube walls detached to enter the water and form there an impure silica gel (quartz is a form of silica). In the case of a glass capillary, various contaminants of the glass – for example, left there during the heat-process of pulling the tube – became dissolved in and/or combined with the water. The frequently repeated calumny that among the fatty compounds found in polywater was sweat from the experimenters' fingers has its origins in the fact that the spectra of "glass-based" polywater were not dissimilar to those of human sweat.

In a tightly restricted sense, polywater did exist: the substance whose properties and behaviour all the various chemists investigated was not an illusion. Their shared error was, rather, in being insufficiently critical of explanations – notably by Derjaguin, eager to capitalize on a phenomenal discovery – as to what the substance actually was. As soon as more rigorous analysis was brought to focus, the truth emerged. This should have happened some years earlier.

At the time, polywater was not merely an arcane chemical concern. There were fears among some of the scientists involved that, should polywater accidentally be let loose in the wider world, all of the world's ordinary water would "flip" into this form – a terminal catastrophe for all life on earth. But there was a bright side, too. As an excited New York Times commented:

> A few years from now, living-room furniture may be made out of water. The antifreeze in cars may be water. And overcoats may be rainproofed with water.

ARGENTAURUM

If the aim of alchemy was in crudest terms to turn base metals into gold, it seemed to have been achieved by the UK-born US chemist Stephen Henry Emmens in 1897. He looked at the Periodic Table of the Elements and convinced himself and thousands of investors, the US Government included, that there should be an intermediate substance between silver and gold that could be easily converted into either of them. He derived this mysterious element, argentaurum, from silver, and was then able to transmute it on into gold which, he claimed, satisfied "every test to which the US Government Assay Office subjects the gold offered to them for sale". The UK physicist Sir William Crookes, with Emmens's permission, repeated the experiment, but his results were ambiguous; the two men did not get on, and the growing friction between them makes it difficult for us now to establish exactly what happened. Their correspondence was reproduced as part of a book Emmens wrote, *Argentaurana, or Some Contributions to the History of Science* (1899).

Was Emmens, who also published "original ideas" on the mutability of gravity, a fraud out to make a quick buck (no pun intended) or a genuine seeker after scientific truth who managed to delude himself and others about what he'd achieved? Unfortunately, he died soon after the turn of the century and the company he founded to derive gold from silver seems to have died not much later, with all records being lost.

MITOGENETIC RAYS

In 1923 the Ukrainian physicist Alexander Gurwitch, Gurvich or Gurwitsch (1874–1954) reported observing the behaviour of curious short-wavelength (ultraviolet) electromagnetic rays emitted by living matter: they transmitted happily through quartz but were blocked by glass. The effect of these rays seemed to be to communicate some life-promoting energy from one organism to the next. Since Gurwitch's original experiment appeared – and still appears – to be sound, and since other researchers conducted experiments which likewise seemed explicable only in terms of his mitogenetic (or mitogenic) rays, it was no wonder the rays' existence was generally accepted throughout most of the scientific community for much of the 1920s.

The notion of mitogenetic rays did not spring from nowhere. It was widely thought at the time that the processes of an organism's development, such as that of the embryo in the womb, would forever be beyond the capacity of physics or chemistry to comprehend. Thanks to a period spent in Germany with pioneering developmental biologist Wilhelm Roux (1850–1924), Gurwitch begged to differ – and quite rightly. He came to believe organic development must be under the control of a "supercellular ordering factor", a sort of energy field that instructed each newly forming cell as to its correct place in the developing organism. It should be possible, he reasoned, to detect this field. The energy for the field must come as a sort of by-product of ordinary metabolism, and the best source for it would therefore be a mass of growing cells. While the main purpose of the energy they produced would be to guide the growth of further cells within their own organism, it ought to be possible to use it to stimulate growth in a different organism, should the second be placed very close to the first.

In Gurwitch's initial experiment, putting the tip of a growing onion root alongside the tip of another onion root seemed to promote cell division in the target root, the effect dropping off with distance from the point of near-contact. Other researchers did equivalent experiments with yeast, bacteria, etc., with similar results. Soon studies extended to the larger scale. Injecting cancer cells into a healthy animal, it was found, would reduce the animal's effusion of mitogenetic rays, although the tumours themselves would continue to be strong emitters even if the animal were killed. A rabbit's eyes were a good source of mitogenetic rays but, if the animal were starved, the intensity declined; after a few days' further starvation, however, the emission of mitogenetic rays would start up again, although now with a different wavelength, reflecting the different metabolic processes at work. The blood of children suffering from Vitamin D deficiency displayed lesser ray-emission than other children's. And so on.

Nevertheless, attempts to measure or even detect the rays by purely physical means – such as photoelectric cells – came to nothing. By the early 1930s, more and more researchers were beginning to call into question the very existence of the rays. In 1935, reviewing the 500 or so papers published on the subject, A. Hollaender and W.D. Claus concluded in the *Journal of the Optical Society of America* that the whole mitogenetic-ray scenario was the product of badly executed science at every level: researchers were looking for results so minuscule as to be statistically insignificant, and of course were finding what they expected to find.

Mitogenetic rays faded from the journal pages soon thereafter, though with the occasional scattered exception right up until the 1960s. Just to confuse the picture, chemical reactions in living cells can emit tiny amounts of electromagnetic radiation – but as visible light, not ultraviolet. It's feasible that some of the original experiments detected this effect; later claims were, however, products of mob delusion.

It's not too much of a stretch of the imagination to relate Gurwitch's mitogenetic rays to the much later fad for Kirlian photography, whereby people deceived themselves into believing they could take photographs of an organism's energy field, or aura. In 1939 the Soviet hospital technician Semyon Kirlian noticed, while observing a patient receiving treatment from a high-frequency generator, tiny flashes in glass electrodes as they were brought close to the patient's skin. Fascinated, he and his wife Valentina experimented by taking photographs of human skin placed in a powerful electric field. Their first target was Kirlian's hand. When photographed this way, it seemed to have a glowing aura. The fad for Kirlian photography didn't start until some decades later, however, when the experiment was recounted in the book *Psychic Discoveries Behind the Iron Curtain* (1970) by Sheila Ostrander and Lynn Schroeder.* Even though many researchers were unable to reproduce the results, claims for Kirlian photography swiftly grew: the aura of a leaf would retain its shape even if you tore off part of the leaf, the aura of someone who'd just drunk a glass of vodka was brighter than before, and so forth. While there was no disputing that living organisms photographed in such circumstances showed a glow, this was soon explained by scientists as being merely the product of a well recognized and understood process called corona discharge. As for the claims that the aura was an "etheric body" whose brightness varied with the well-being of the organism, John Taylor reported in his confessional *Science and the Supernatural* (1980):

> When all of these factors are carefully controlled, there is no change of the Kirlian photograph [of a fingertip] with the psychological state of the subject. It seems that the most important variable is the moisture content of the fingertip, and the Drexel group even concluded that "corona discharge (Kirlian) photography may be useful in detection and quantification of moisture in animate and inanimate specimens through the orderly modulation of the image due to various levels of moisture."

Such a mundane conclusion was of course ignored by the media and the public, who preferred to think the glow around the vodka drinker's hand brightened not because he was sweating a little more but because he was feeling cheerier.

* Quite why a perfectly ordinary scientific experiment should be regarded as a "psychic discovery" is anyone's guess. Whatever the source of the glow seen in Kirlian photographs, there seems nothing especially paranormal involved.

MENSTRUAL RAYS

Amid all the accounts of various rays that have later proved illusory, one particular ray-related phenomenon has gone underreported. A considerably body of popular fallacy surrounds menstruation. In some parts of Australia the Aborigines maintain that, if a man goes near a menstruating woman, he will lose his strength and grow prematurely old. Similar myths are alive in Western society. Eating ice cream during one's period is risky – but not as bad as washing one's hair. Bathing during a period can lead to tuberculosis. Flowers wilt and crops are blighted as menstruating women walk by.

That there might be more to this latter superstition than just, well, superstition occurred to a German bacteriologist called Christiansen in 1929. He noticed that the fermentation starter cultures in his lab failed periodically, and linked these failures to the menstrual cycle of the female technician who tended them. Clearly either menstrual blood or menstruation itself was a source of radiation, and like a good scientist Christiansen investigated. The properties of menstrual rays proved to be not unlike those of Gurwitch's mitogenetic rays – they passed through quartz but were blocked by glass, etc. – but their effects on developing organisms were negative where those of mitogenetic rays were positive.

Similar effects were noticed in the US, at Cornell University, by the bacteriologist and mitogenetic-ray enthusiast Otto Rahn (b1881), in whose laboratory yeast cultures had monthly difficulties when tended by a particular female student. As had been observed by Christiansen, the effect was seasonal: in winter it was barely noticeable, while in summer it was at its strongest. Rahn, too, conducted investigations, and concluded that menstrual radiation was just the most obvious case of a more generalized radiation human bodies could give off in certain circumstances, such as hypothyroidism, herpes and sinus infection. Rahn went on to publish a whole book on the subject, *Invisible Radiation of Organisms* (1936). One hardly need add that the topic isn't much discussed by scientists today.

A MICROBIAL MUDDLE

We noted briefly how the early microscopists were deceived by poor optical equipment and wishful thinking into seeing things that weren't there. They haven't been the only ones. According to G.M. Boshyan in his *On the Nature of Viruses and Microbes* (1949), something of a bestseller in Moscow, viruses and microbes are really one and the same thing, transmutable into each other. Boshyan was a student at the Moscow Veterinary Institute, and he'd made a most remarkable discovery while peering through his microscope. When he treated microbes by boiling or chemicals they did not die: instead they were transformed, depending upon circumstances, into either viruses or bacteria. With further treatment, clusters of viruses could be induced into massing together to form a bacterium, or a bacterium could be persuaded to break down into its constituent viruses.

Anywhere else in the world, Boshyan might have been recommended to take his researches in other directions, but this was the USSR under the sway of Lysenkoism (see pages 270ff), one of whose bases was that environmental factors could persuade species to transform one into another. Boshyan was declared to have made a major breakthrough in molecular biology and given a large laboratory of his own; further, the discoveries were of such importance that they should be classified in case they were stolen by hostile foreign powers.

Despite the dangers of disputing the official line on the biological sciences during Lysenko's sway, the Soviet Academy of Medical Sciences decided to investigate. For a long while their investigative panel was stalled by the classified nature of Boshyan's work. When, finally, they were permitted a demonstration, their unequivocal conclusion was that Boshyan had failed to clean his microscope slides properly. For once, reason prevailed: Boshyan's entire theory swiftly unraveled, and he was thrown out of the Institute. The case seems to have been one of synergy: indoctrinated by both Marxism and Lysenkoism, Boshyan saw what he wanted to see, which was also what his masters would like him to see, so he saw it even more emphatically . . .

TALKING APES

In 1979 psychologist Herbert S. Terrace (b1936) pricked
the bubble of the seemingly thriving research into the ability of higher apes to comprehend and use language via the
American Sign Language (ASL, or Ameslan), the gesture
equivalent of spoken language as deployed by the human
deaf or dumb. (Apes lack the vocal cords that would allow
them to speak human language.) Terrace had conducted an
experiment with the chimpanzee Nim Chimpsky
(1973–2000); the chimp, who had learned about 120
gestures, seemed to use them syntactically to construct
sentences. Soon after the conclusion of the experiment,
however, Terrace had doubts about his own conclusions. Re-
examining Nim's behaviour on video tape, he saw that Nim
was not constructing sentences, as he had thought, but
merely imitating his tutors. Further, Terrace concluded the
same was true of all other such experiments with the apes,
of which several were ongoing. A conference was held in
1980 under the auspices of the New York Academy of
Sciences, "The Clever Hans Phenomenon: Communication
with Horses, Whales, Apes, and People",* and the whole
field of research was trashed by the participants – to the
delight of the media, who gave the affair blanket coverage.
     The full picture wasn't quite as simple. Terrace, a
Skinnerian behaviourist, had taught Nim according to the
principles of that school of psychology: rather than, as other
researchers were doing, raise the chimp in a manner analogous to the raising of a human child, with language being
brought naturally into the mix, Terrace taught Nim in a
small, featureless room for two three-hour sessions each
weekday, with up to 60 different tutors substituting for the
"parents" in the other experiments. Further, Terrace used
no double-blind to obviate the chimp learning merely to
imitate his tutors in order to gain rewards: this was the
Clever Hans Effect with a vengeance. When Nim was later
introduced to a community of other ASL-using chimps he

* The Clever Hans Effect is named after a famous horse who was
believed, early in the 20th century, to be able to count: a number being
given to him, he would tap the ground with his hoof that number of
times. It was proved the horse had learned to read the unconscious
movements of his trainer as an indication of when to stop tapping.

soon started to pick up the use of language. All that the Nim experiment had really proven was that young apes – like young children – don't learn according to Skinnerian principles.

The most famous of the ape-language experiments has been that of Allen and Beatrice Gardner of the University of Nevada, working first with the female chimpanzee Washoe (b1965) and then with others. There is still some controversy about the results of the Washoe experiment, although the balance of opinion is that she did indeed learn how to "talk". Other experiments, notably with the pygmy chimpanzee Kanzi (b1980) and the gorilla Koko (b1971), seem more than adequately to have replicated the Washoe findings. Interestingly, Koko has over the years "adopted" several cats as pets; she was the model for the talking ape in Michael Crichton's novel *Congo* (1980).

Nim Chimpsky

RECOVERED MEMORIES

No one could doubt child sexual abuse is an abhorrent crime, and that its incidence is far more widespread than any of us realized before the spotlight was turned on it. At the same time, it seems far *less* prevalent than was generally reported during the height of the hysteria in the 1980s and 1990s, when claims were based on supposed science involving the recovery of repressed memories – which memories often linked the sexual abuse to the parents' participation in Satanic rites. The notion was that many children found the experience of sexual abuse so awful that they dissociated themselves so comprehensively from the act as effectively to become multiple personalities; later, they had no memory of the abuse because it happened to "someone else". They did, however, display a complex of symptoms – lacklustre eyes, inappropriate responses to certain triggers (especially sexual ones), and so forth – and these could be read by the wise therapist. With encouragement, the patient might be able to retrieve the memories that had been "lost", and from there it was but a short step to the criminal conviction of the accused abuser and the recovery of the patient through "closure".*

Hypnosis was a popular tool for the recovery of the memories, and many an innocent person was jailed before it became recognized that all the hypnotic subjects were doing, like hypnotic subjects in any other context, was telling the hypnotist what they thought s/he wanted to hear – reflecting back the therapist's own preconceptions, in other words. What those preconceptions were was only too easy for the hypnotized patient to deduce from the questions themselves. Even after hypnosis had been abandoned as a tool, and a fraction of the falsely imprisoned had been released,† the recovered-memory movement continued unabated, a political rollercoaster powered by unlikely allies such as militant feminists and the religious right (thrilled by the Satanism angle).

Yet from the very outset there has been little scientific evidence that repressed memories exist at all. To be sure, victims of concussion typically have a "blank period" prior to the trauma: it may be a long while, if ever, before fragmentary memories return of the minutes or hours leading up to the blow. Abrupt psychological trauma can similarly engender a "blank period". But these are very, very different processes from a pattern of repeated sexual abuse and its usual accompaniment of threats, either of violence or to "tell everyone you're a dirty little girl/boy who seduced me, so that no one will love you"; even if the model of memory-loss through dissociation during the act itself could be accepted, it is hard to see how the rest could be forgotten. It is today the overwhelming consensus of scientific opinion that repressed-memory syndrome is a fiction, yet this, while it has slowed the political juggernaut, has failed to stop it. As late as 2005 we have the (otherwise very good) novel *Blood Money* by Greg Iles swallowing the pseudoscience of recovered memories wholesale – indeed, being based

---

* Itself a deeply unscientific concept.
† Some still rot in jail, pariahs even there because of a crime they did not commit.

almost entirely upon it. In his "factual" Afterword, Iles makes the very fair point that we should never, ever ignore those who claim to be the victims of sexual abuse. What he fails to point out is that neither should we believe those claims simply because they've been made. The parallel with the 1692–3 case of the Salem "witches" is obvious.

It's important here to recognize that the falsely imprisoned were not the only victims of the recovered-memory cult. Countless people, usually women, came to have their lives haunted by "memories" of appalling acts that had never been committed. No one can assess the psychological damage this has done them. And likewise no one can assess the damage done by the failure, through focusing on fallacious "memories" of abuse, to address whatever was the *real* problem that brought them into therapy in the first place.

Nor should we forget that the techniques of the repressed-memory therapists have elicited similarly convincing accounts of people's abductions by UFOs.

PRAYER POWER

The prayers of others can make a woman pregnant! No, no, this is not your daughter's despairing excuse; it's according to a paper published in 2001 by the *Journal of Reproductive Medicine*: "Does Prayer Influence the Success of In Vitro Fertilization-Embryo Transfer?" by Kwang-Yul Cha of the Cha Hospital, Seoul, and the Cha Fertility Center, Los Angeles, and Rogerio A. Lobo and Daniel P. Wirth of Columbia University, New York. They reported an experiment in which 199 women who required in vitro fertilization were separated into two experimental groups. Without the subjects' knowledge, numbers of volunteer Christians in the US, Canada and Australia prayed over photographs of the women in one group for a successful fertilization. The women in the other group were left unprayed-for as a control.

The results reported in the *Journal* were impressive. Of the prayed-for women, 50% attained a successful pregnancy, while of the others only 26% did so (which latter percentage is about the standard rate for in vitro fertilization attempts). According to Lobo, given as the paper's senior author, the three scientists spent some time considering whether or not they should publish their results, because the difference in pregnancy rates was so improbably dramatic – too good to be true.

Others reached that same conclusion. Even while newspaper headlines and evangelists were shouting worldwide about proof of the power of prayer, researchers began investigating the paper's claims. One such, Bruce L. Flamm, Clinical Professor at the Department of Obstetrics and Gynecology, University of California Irvine Medical Center, published an extensive three-part analysis in 2003–2004 in the *Scientific Review of Alternative Medicine*. He found much troubling in the experiment's methodology and conclusions – and indeed in the modern trend for peer-reviewed journals to publish, without seemingly doing much by way of peer-reviewing, papers that invoke the spiritual or supernatural: "It is one thing to tell an audience at a tent revival that prayers yield miracle cures, but it is quite another to make the same claim in a scientific journal." A further worrying fact was that one of the paper's three authors, Wirth, had published extensively in support of the supernatural; was it reasonable to expect his contribution to be unbiased?

Drawn into the spotlight, Lobo began wilting. Columbia University issued a statement on his behalf saying that the first he had known of the study was some 6–12 months after its completion; he had merely provided "editorial assistance" for the paper . . . and allowed himself to be listed as senior author. And then in 2002 coauthor Daniel P. Wirth – who had given his affiliation here and in other paranormal-supporting papers as "Healing Sciences Research International", an institution in California which he headed and which seems to have consisted primarily of a PO box – was indicted for large-scale fraud. In April 2004 he and Josepf Steven Horvath pleaded guilty to conspiracy to commit mail and bank fraud, and agreed to forfeit assets totalling over $1 million gained by various schemes. In this context, with Cha, like (by now) Lobo, unwilling to return phone calls or e-mails, it seems legitimate to speculate that they were as much taken in as anyone else by Wirth, and that the results of the experiment were "manipulated" – if indeed the experiment ever took place at all.

The Lobo *et al.* study has not been the only controversial paper on the purported therapeutic effects of prayer power to be published in a peer-reviewed medical journal. Mitchell Krucoff, a cardiologist at Duke University Medical Center in Durham, NC, and his team focused on the therapeutic effects on patients suffering congested coronary arteries of prayers offered up on their behalf by people unknown to them. A pilot study was published in 2001 in *American Heart Journal*, with a fuller study in *Lancet* in 2005. In the wake of the 2001 paper Krucoff was interviewed about his exciting results on the Discovery TV channel, and summarized:

> We saw impressive reductions in all of the negative outcomes – the bad outcomes that were measured in the study. What we look for routinely in cardiology trials are outcomes such as death, a heart attack, or the lungs filling with water – what we call congestive heart failure – in patients who are treated in the course of these problems. In the group randomly assigned to prayer therapy, there was a 50% reduction in all complications and a 100% reduction in major complications.

He made similar startling claims elsewhere. However, Andrew Skolnick, Executive Director of the Commission for Scientific Medicine and Mental Health, took the trouble to compare Krucoff's public statements with the results recorded in the *American Heart Journal* paper itself, and discovered that to say "In the group randomly assigned to prayer therapy, there was a 50% reduction in all complications and a 100% reduction in major complications" was not entirely accurate: just for a start, while all of the control group had survived until the end of the experiment, one of the prayed-for group had died – definitely a "major complication". Pressing Krucoff on this point in a 2004 interview, Skolnick received the reply: "Well, the difference between zero and one in a cohort of 30 people is no difference." Even if one could accept that, what it most certainly isn't is a *reduction*.

The more complete study published in the *Lancet* showed, as one might expect, absolutely no effect one way or the other of prayer on the patients. However, this is not the impression you might get from the press release issued on July 14 2005 by Duke University or the *Lancet*'s editorial, both of which imply that the results are, for some unexplained reason, hopeful. In the press release Krucoff says:

> While it's clear there was no measurable impact on the primary composite endpoints of this study, the trends and behavior of pre-specified secondary outcome measures suggest treatment effects that can be taken pretty seriously when considering future study directions.

Translated into English, this means: "We found no effects whatsoever, but we're going to try again." As presumably Krucoff will. Inevitably, if a number of studies are done, there'll finally be one which, purely by chance, appears to show the prayed-for patients doing better than the control group. One of the weaknesses of the current scientific process is that, as has often been noted, "successful" studies are usually the ones published, and everybody forgets about the others.

In March/April 2006 the *American Heart Journal* published another paper, done by Herbert Benson (a cardiologist from the Mind/Body Medical Institute in Massachusetts) *et al.*, reporting on the effects of prayer on 1800+ patients recovering from heart surgery, an experiment conducted over more than a decade. Heralded as the most scientifically rigorous study in the field to date, this found that the prayers of strangers had zero effect on the patients' recovery. Perhaps counterintuitively, when patients knew they were being prayed for they had higher rather than lower rates of post-operative complications; Benson and his colleagues suspected that the knowledge of the prayers inappropriately raised the patients' expectations for a rapid recovery. In other words, prayer power's *bad* for you!

Despite the very public discrediting of the earlier reports, the belief in the curative power of prayer seems to be deeply ingrained. In early 2006 it was reported that the Church of Christ, Scientist, was gearing itself up to resist the imminent threat of a bird flu pandemic by use of prayer – and not necessarily only with other Christian Scientists: apparently, *in extremis*, anyone would do. This was in accordance with the Christian Science doctrine that disease is a spiritual rather than a material phenomenon, its undeniable physiological effects being due to the fear of the sufferer: according to Mary Baker Eddy's *Science and Health*, "Disease is an experience of so-called mortal mind. It is fear made manifest on the body."

In late 2005 the televangelist Darlene Bishop was sued by the four children of her late brother, the songwriter Darrell "Wayne" Perry (d2004), for having persuaded him to give up the course of chemotherapy he was undergoing for his throat cancer and to rely instead on prayer and God's healing powers.

CLINICAL ECOLOGY

Clinical ecology is a "science" that's taken seriously in courtrooms but less so outside them. The initial premise of clinical ecology is not absurd. In *Dirty Medicine* (1993), Martin J. Walker sums it up thus: "The philosophy and practice of what is generically called clinical ecology assumes that the mechanical and chemical processes of the Industrial Revolution, the electrical and the nuclear age, have all had a deleterious effect upon the health of individuals and societies." There is much to applaud here: it is thanks in part to the efforts of clinical ecologists that we have become aware of the dangers of lead in our gasoline and mercury in our fish. One can even sympathize with Walker's general sentiment when he says: "The basic demand of clinical ecology is . . . a radical one: that the industrial means of production be reorganized to suit the health of the whole of society."

Where clinical ecology falls down is that it soon becomes possible to attribute just about any ailment to just about any element of the environment, which opens up a vast ballroom in which the litigation lawyer and the sham "expert witness" can dance. And the dubious science purveyed in the courtroom for reasons of financial gain inevitably diffuses into the public consciousness, where it all too easily becomes "accepted scientific fact". Thus we've had scares that proximity to high-voltage electrical cables and junction boxes causes cancer: there is no evidence whatsoever of this. We've had scares about atmospheric levels of radon so low as to be positively homoeopathic. We've had scares about "chemical AIDS", which supposedly involves chemicals in the environment suppressing the immune system . . . except that they most demonstrably don't. And so on and so on.

The big legal triumph of clinical ecology, according to Peter Huber in his book *Galileo's Revenge* (1991), came in 1985 in the town of Sedalia, Missouri, where the local chemical plant, Alcolac, was blamed for pollution that caused a number of the residents to suffer "chemical AIDS". Appearing for the plaintiff as medical experts were Bertram W. Carnow, who "registered for the board certification exam in internal medicine in 1957, 1958, 1960, 1961, 1962, 1963, and 1964, but withdrew twice and failed five times", and Arthur Zahalsky, an immunologist who "never actually studied immunology in graduate school; but he does claim to have audited immunology classes at Washington University in St Louis".

The field of litigation is in general rife for scientific corruption. As Huber points out from a lawyer's perspective, "[A]s you labor to assemble your case, the strength of the scientific support for an expert's position is quite secondary. It is the strength of the expert's support for *your* position that comes first." In other words, who cares if your "expert witness" is a pseudo- or fringe scientist whose testimony is junk science or plain lies so long as it wins you your case?

And junk science such testimony often is. It is surprising the laws to curtail perjury are not more often invoked: any other witness who presented an extreme implausibility as if it were an established fact would at the very least be cautioned by the judge, but there's crept into the court system the notion, stated or unstated, that scientific fact – or scientific consensus, which is almost synonymous – is somehow open to debate, or even democratic vote.*

There are other factors involved. Science in the real world is assessed by the community of other, relevantly qualified scientists; in the courtroom it is only by chance if a member of a jury has the knowledge to dissect what an "expert witness" has testified. Generally it is laypeople

---

* This is a corruption that plagues modern Western society in general, as we shall see later.

who're trying to cope.* Plausibility becomes everything; truth becomes immaterial. Similarly, whereas any competent scientist will hedge any statement of presumed fact with qualifications – "so far as is understood", "the general consensus of opinion" and so on – this tentativeness, which in science is a sign that someone knows what they're talking about, is all too often regarded by the layperson as the opposite: the purveyor of the definitive statement is regarded as the more authoritative witness, the more plausible "expert".

The biggest scientific corruption of all in connection with the legal system arises because "expert witnesses" are paid if they agree with the lawyer who wants to hire them, and not paid if they don't. The pay is often quite handsome – to the extent that many second-rate scientists find it all too tempting to abandon their academic careers for the riches offered by a career as a serial "expert".

Corrupt science in court affects us all. A couple of decades ago successful lawsuits against the manufacturers of whooping-cough vaccine drove millions of parents, quite reasonably, to reject it for their offspring on the grounds that it could cause all sorts of disabilities, up to an including brain damage and death. The truth was that, since vaccines work by inducing mild forms of the disease against which they protect, there was the most remote chance that some child, somewhere, might indeed be affected by whooping-cough symptoms as a result of being given the vaccine. There was no such instance on record, and numerous large-scale clinical trials had failed to turn up even the slightest trace of such an effect, but it was not completely outwith the bounds of possibility. That was enough for certain courts, prompted by "expert witnesses", to rule that the vaccine was at fault in a few tragic cases . . . and the panic began. Of course, the full-scale disease *does* produce those symptoms, and infants suffer life-wrecking damage and death because of them; no one has yet assessed how

* And sometimes they cope very badly indeed. In the 1943 paternity suit brought by Joan Berry against Charlie Chaplin, Chaplin's lawyers presented definitive scientific evidence that the actor could not have been the father of Berry's child: blood-typing proved it impossible. Nonetheless, the jury decided in Berry's favour.

many deaths were caused by the corrupt science embraced by the courts in this instance, but a conservative guess is thousands. Similarly, the morning-sickness drug Debendox, also known as Bendectin, was driven off the market in the US because of a few high-profile judgements based on fake evidence (see page 49). Morning sickness, although mothers tend afterwards to joke about it, is in its worst cases seriously debilitating, extremely unpleasant and indeed life-threatening to the mother, and can cause miscarriage. Again, no one has attempted to calculate how many mothers and their unborn offspring have died because of the court decisions, but it's a statistical certainty that there've been quite a number, not to mention the countless pregnant women who've suffered unnecessary misery.

To stress: A court of law is not the place to decide a point of science, yet all too often this is in effect what happens. It's a point we'll return to when we look at the tangled course of the argument over teaching Creationism in US schools (see pages 155ff and 165ff).

HOT STUFF IN THE COURTROOM

One legal field in which science has been extensively corrupted through an overwillingness by prosecutors and investigators alike to rely on the lessons of experience, even where the experience itself rests on little more than years of accumulated, self-reinforcing guesswork, often faulty, is arson investigation. In a number of instances, unfortunates have, based on evidence presented as scientific but in fact little more than wishful thinking, been executed as murderers for setting fatal fires that have since been identified as accidental.

One such was Cameron Todd Willingham, executed in Texas in 2004 for setting the fire that killed his three children in Corsicana, near Dallas, in 1991. At Willingham's trial the Deputy State Fire Marshal, Manuel Vasquez, presented damning evidence of the fire having been artificially set. A US group called the Innocence Project, founded in 1992 to reinvestigate dubious convictions through proper scientific analysis, commissioned a panel of five arson experts to sift through the evidence in Willingham's case; even before Willingham's execution, some of this evidence was presented to Texas Governor Rick Perry (b1950), who ignored it. The conclusion of the panel was that the fire had been accidental; there was no reason whatsoever to doubt Willingham's own account. The "scientific" evidence that had convicted him represented Vasquez's no doubt sincerely held opinion, but that opinion, which had persuaded the jury to convict, was based not on science but on, essentially, folklore.

One of the items of evidence Vasquez presented was the way that some of the glass in the Willingham home had been crazed; this, he told the court, was a clear indication of the high temperatures generated by fire accelerants, and could have been caused in no other way. Such indeed was the conventional wisdom among arson investigators at the time, but it was based on a complete lack of scientific experimentation. Two years later, in 1993, using the aftermath of the 1991 wildfire in Oakland, California, as their testing ground,* a team of investigators including John Lentini, later one of the Innocence Project's five experts in the Willingham case, put a deal of arson's conventional wisdom to the test. They found crazed glass aplenty, and laboratory follow-up work showed that no amount of heating would produce the "telltale" crazing; what produced the crazing was rapid *cooling*, as might occur when the glass was hit by the water from a fireman's hose. Decades of conventional wisdom on the matter had been demolished. Vasquez could not have known this; the results of the work by Lentini *et al.* were, however, publicly available a full decade before Willingham's execution.

Eight months after Willingham's death, Texas exonerated another man, Ernest R. Willis, who on similarly fallacious evidence had been convicted and sentenced to death for killing two women in a house fire in 1986 in Iraan, Texas. For his 17 years of wrongful imprisonment Willis collected $430,000 in compensation from the state. The same "new" (in fact, decade-old) science that exonerated Willis had been rejected in Willingham's case.

The 1993 team at Oakland made various other ground-

* The advantage of the Oakland site, which contained some 3000 destroyed homes, was that, whatever the cause of the original fire (in fact, natural), there was no question but that homes away from the epicentre had been ignited naturally.

breaking scientific discoveries. Another indicator that had been used for years as evidence of high temperatures, and hence of the use of accelerants, was melted steel – typically melted bedsprings. The team found numerous examples of bedsprings that appeared to have melted but which, when examined in the laboratory, proved instead to have suffered extensive oxidation, which happens at far lower temperatures. A more reliable indicator of high (albeit somewhat lower) temperatures remained melted copper; the patterns within the home of the melting, something into which investigators had for a long time read much, was, however, random, as were the patterns with which steel objects like bedsprings oxidized.*

Some while later, Lentini got involved after the conviction of Han Tak Lee of having murdered his mentally ill daughter by setting a fire at a religious camp in Stroud, Pennsylvania. In this instance the "scientific" evidence presented by the state was a mixture of disproven conventional wisdom and straightforward hokum; the principal "expert", Daniel Aston, a part-time suspicious-fire investigator, claimed to have examined some 15,000 fires, whereas the busiest full-time investigator might be expected to examine something under 5000 fires in the space of an entire career. Nonetheless, Aston's claim went unchallenged by the defence, the same going for the astonishingly detailed conclusions he presented. Lentini's own conclusions, given in his paper "A Calculated Arson",† are worth quoting:

> The quality of the evidence presented by the Commonwealth [of Pennsylvania] speaks for itself. Fuel loads calculated to six significant figures, hydrocarbon "ranges" being interpreted as evidence of a mixture, furnace operating instructions being touted as normal fire behavior, and a host of other "old wives tales" were used to convict Han Tak Lee.
>
> There are over 500,000 structure fires every year in the

* "Unconventional Wisdom: The Lessons of Oakland" by John J. Lentini, David M. Smith and Richard W. Henderson, *Fire & Arson Investigator*, June 1993.
† *Fire & Arson Investigator*, April 1999.

United States. (Approximately 15% are labeled suspicious or incendiary.) Each presents an opportunity for erroneous cause determination, and a significant number of erroneous determinations do occur. Even if fire investigators are correct 95% of the time, that allows for 3,000 incorrect determinations of arson each year. These calculations demonstrate the need for objective investigations based on the scientific method.

Much has been written lately about the criminal justice system allowing guilty people to escape justice due to sloppy police work. Here is the case of the wrongful conviction of an innocent man, surely a worse result. The Lee case represents the ultimate triumph [in the courtroom] of junk science.

If arson investigation were the only area in which forensic science was in chaos, that might seem a containable problem – one that could be dealt with by stepping up the scientific training of arson investigators, many of whom have received none at all: they have learned the principles of their trade through training on the job under other, more experienced investigators who've unwittingly passed on dubious information. But a flurry of recent investigations have shown that much forensic science in other areas is of equally unsound basis, including even that old staple, fingerprints: it is not quite true that every fingerprint is unique, while nuances of interpretation in fingerprint analysis can all too easily lead to false identification. Throughout, even in the supposedly infallible DNA testing, forensic science is like everything else prone to corruption through simple human error, while personal traits such as arrogance among forensic scientists may also play their part. The fervour of prosecutors eager to obtain a conviction at any price, including that of the truth; the willingness of even the best-intentioned juries to see their biases confirmed; and the readiness of junk scientists and outright pseudoscientists to bask in the glory of accolades as expert witnesses in the courtroom – all of these corrupt the science yet further.

Above: What we did to Hiroshima with an A-bomb
Backdrop: The H-bomb

CHAPTER 3

MILITARY MADNESS

———————<✦>———————

Despite the vision and the far-seeing wisdom of our wartime
heads of state, the physicists felt a peculiarly intimate responsibility for suggesting, for supporting and, in the end, in large
measure, for achieving the realization of atomic weapons. Nor
can we forget that these weapons, as they were in fact used,
dramatized so mercilessly the inhumanity and evil of modern
war. In some sort of crude sense which no vulgarity, no humor,
no overstatement can quite extinguish, the physicists have
known sin; and this is a knowledge which they cannot lose.
    J. Robert Oppenheimer,
    "Physics in the Contemporary World",
    lecture delivered at MIT, 1947

The practical use of guided missiles can only be to kill foreign
civilians indiscriminately, and to furnish no protection whatsoever to civilians in this country. I cannot conceive a situation in
which such weapons can produce any effect other than extending the kamikaze way of fighting to whole nations.
    Norbert Wiener,
    "A Scientist Rebels", *Atlantic Monthly*, January 1947

In the councils of government we must guard against the
acquisition of unwarranted influence, whether sought or
unsought, by the military-industrial complex. The potential
for the disastrous rise of misplaced powers exists and will
persist.
    President Dwight D. Eisenhower,
    Farewell Address to the Nation, January 17 1961

A GOOD CASE CAN BE MADE THAT, whenever science is
controlled by nonscientists, the result is a corruption of
science. Yet, in our modern world, science almost always *is*
controlled by nonscientists, sometimes by political edict but
most significantly though selective funding – the allocation
of money for research by governmental, commercial or military organizations. In 2001 in the US, even before the start
of the "war on terror", an annual federal research budget of

some $75 billion saw nearly $40 billion go to the Pentagon, primarily for research into weaponry, while a mere $4 billion went to the National Science Foundation for actual *science*. That disproportion is in itself an obscene corruption.

Yet consider the allocation of its vast budget by the Pentagon, with decisions being made by nonscientists as to which lines of research should be pursued, which discarded. To any criticism of the situation the reply is that, obviously, the military should decide which weapons show most promise. Yet is that so "obvious"? Turning the argument around, would the military think it reasonable if the scientist who'd created a new weapon for them then dictated the strategy for its deployment? Well, no: the military would regard that as wholly unreasonable – and they'd be right.

The meddling in science by the Nazi hierarchy, through their support of pseudoscience and through their imposition of a daft antisemitic ideology on the structures of scientific research, inadvertently went a long way toward ensuring that the Reich never developed nuclear weapons; for that we must be thankful. Yet, in the decades since the destruction of the Reich, governments of all stripes, and their military institutions, have failed to recognize the lesson: untrained personnel are ill equipped to dictate the course of scientific research. Just as a politician or a religious demagogue – or a journalist – lacks the tools to judge the desirability (or otherwise) of stem cell research, so a soldier is incapable of discriminating between a sane line of weapons research and one that's loony.

In this chapter most of the madness in the field of military technology we look at is from the post-war US, simply because that nation has been by far and away the biggest spender in military R&D, and therefore has seen the least necessity to trim the relevant budgets; indeed, any US politician who suggests a cut in spending in this area is likely soon to be an ex-politician. It is by popular vote that the enormous waste of public money on crazily corrupted science, or outright pseudoscience, continues. Quite how significant the waste is can be exemplified by the fact that, in the years after 2003, an Iraqi resistance whose typical weaponry was the home-made bomb could fight to a standstill the most technologically advanced army in the world.

The propensity for the Pentagon to spend colossal amounts of money on unnecessary technology, all without proper oversight by the taxpayer, perhaps first became fully evident to the US public in the 1960s and 1970s, with the C–5A Galaxy Scandal. This enormous military transport aircraft was commissioned from Lockheed at $20 million per plane; the final cost was just under $60 million per plane. This represented a total cost for the project that exceeded the annual budgets of Congress and the US Departments of State, Justice, Interior and Commerce combined. It was a remarkable aircraft in theory: it could land on runways less than six times its own (prodigious) length, when on the ground could raise and lower itself for ease of loading and unloading, and so on. Lockheed boasted: "It's a new kind of defense system. It's like having a military base in nearly every strategic spot on the globe." This promise concerned some in the US Government on strategic grounds: how good an idea is it that the world knows you're able and willing to drop two or three divisions into their home patch within a matter of hours?

That aside, the C–5A had abundant technical problems, up to and including wings cracking and wheels dropping off. The problems led to substantial cost overruns as the engineers tried to sort out the technology: for months some $2 billion of this overrun (big money in those days, today a mere bagatelle in terms of the US military's ever more grandiose weapons spending) was concealed from Congress and the public, and when finally someone blew the whistle on it the Air Force's response was to fire the whistleblower, A. Ernest Fitzgerald, the Pentagon Deputy for Management Systems.* Once the plane was in service, there were multiple failures in operation, including several fires and crashes. In the mid-1970s wing cracks were found throughout the entire fleet, so that its permissible cargo weight had to be drastically cut back; it took until 1986 to "re-wing" the planes. By that time, despite all that had gone before, the Air Force had commissioned Lockheed for, and indeed had just begun to accept delivery of, the C–5B . . . It seemed there was no way for Congress, far less the US public, to prevent this haemorrhage of cash on an aircraft that many felt was not just unnecessary but undesirable.

THE STRATEGIC DEFENSE INITIATIVE

One of the most dramatic examples of scientists fooling themselves concerns Edward Teller (1908–2003) and the Strategic Defense Initiative (SDI, or "Star Wars"). There's no doubting Teller was a brilliant physicist; there's also no doubting he was a man of huge ego, to the point that he had difficulty accepting he could ever be wrong about anything of importance. From early in his career it was evident he was one of those scientists whose strength lay in his prolificity of ideas, often astoundingly unorthodox ones; it was left to others to do the detail work, the math, to evaluate those ideas. In the event, some 90% would prove valueless, but the remaining 10% still represented far more *good* ideas than the average theoretician could dream of.

* This was a singularly foolish move. Fitzgerald promptly wrote the book *The High Priests of Waste* (1972) giving the full blow-by-blow insider account of all that had gone on, including corruption within Lockheed, the Pentagon and the US Government, in connection not just with this scandal but with others, such as the F–111 fighter and the Minuteman missile. Fitzgerald also recounted the experiences of a later whistleblower, the former Lockheed staffer Henry Durham, who as he left the company was warned by a Lockheed Director of Manufacturing, Paul Frech, of the consequences of any whistleblowing he might do. Durham: "[Frech] said Fitzgerald would never be able to get a good job as long as he lives. He gave me to understand that anybody who bucks Lockheed or the Air Force is in for a rough time for the rest of his life."

Right: J.Robert Oppenheimer (right). *Below:* Los Alamos during World War II

After he fled the Nazis for the US (via, briefly, the UK) in 1934, Teller's ascent of the scientific ladder was rapid. In due course he became a part of the Manhattan Project at Los Alamos under J. Robert Oppenheimer (1904–1967) to create the atom bomb, although his irascibility saw him progressively sidelined from the project's main thrust. Even before the team's success in producing the A-bomb, which relies on the *fission* of radioactive isotopes of heavy metals, Teller was pushing for the development instead of the H-bomb, which depends for its working on the *fusion* of atoms of hydrogen (or, better, its common isotope deuterium, "heavy hydrogen"). The principal advantages of a fusion bomb are that its fuel is common and cheap and that the energy release is very much greater – considerably more bang for the buck, in other words. Although Teller later rejoiced in the honorific "Father of the Hydrogen Bomb", in fact the scientist whose conceptual breakthrough was primarily responsible for turning the device from a pipe-dream (or pipe-nightmare) into a reality was Teller's colleague Stanislaw M. Ulam (1909–1984); characteristically, in later life Teller was loath to mention Ulam's joint paternity.

In the early 1950s Teller, now at loggerheads with most of the scientific community at Los Alamos, persuaded the powers-that-were in the US Air Force that a second nuclear weapons lab should be founded, with himself as its scientific head. Although Oppenheimer and the Atomic Energy Commission vigorously opposed the move, the Lawrence Livermore National Laboratory, named for Ernest Lawrence (1901–1958), inventor of the cyclotron, was founded in 1952 at Livermore, California. Here Teller initially produced a string of embarrassing flops. He had great difficulty in creating an H-bomb that would actually work: several tests fizzled before finally Teller's team succeeded in creating a detonation, long after the Los Alamos team had conducted their own successful H-bomb tests. By now Teller regarded Oppenheimer as his bitter enemy, and he took advantage of the McCarthy anti-communist witch-hunts to destroy the older scientist, testifying before the Atomic Energy Commission in April–May 1954 to devastating effect. The profoundly right-wing Teller did not say that the moderate Oppenheimer had communist leanings; instead he attacked Oppenheimer's opposition to the H-bomb, leaving others to draw their own conclusions: "I would like to see the vital interests of this country in hands which I understand better, and therefore trust more." Indeed, Teller's testimony to the FBI on this issue is widely thought to have been responsible for initiating the investigation of Oppenheimer in the first place.

In *Forbidden Knowledge* (1996) Roger Shattuck contrasts the introspection of the Oppenheimer remark cited at the head of this chapter with Teller's comment in a 1994 interview concerning the development of the H-bomb: "There is no case where ignorance should be preferred to knowledge – *especially* if the knowledge is terrible." While there is much to agree with in this *per se*, the difference between the moral worldviews of the two men could not be greater, and it was this difference which alienated them from each other and led Teller to seek, and achieve, Oppenheimer's downfall. Oppenheimer saw the shades of grey that make up genuine morality; Teller dealt with the stark blacks and equally stark whites of the *faux*-moralist. It was a view he repeated forcefully in 1988 in Washington DC on meeting Andrei Sakharov (1921–1989), a major contributor to the Soviet H-bomb but later, through his campaigning for a nuclear test-ban, the recipient of the 1975 Nobel Peace Prize. The Soviet dissident emigré denounced the practice of atmospheric testing of the H-bomb, on the grounds that untold thousands of people would be affected by the radiation, and he denounced SDI as a threat to world peace through its demolition of the nuclear balance. Teller's response was that to suppress either would be, aside from impairing the defence of the US, to suppress the advance of human knowledge. Besides, the doses of radiation people might receive from the testing were minuscule, far too small to affect them adversely. In this latter contention he was strictly at odds with the overwhelming conclusions of research, but his ideology blinded him to the light of genuine science.

To return to the 1950s: Although the Livermore laboratory had got its H-bombs to work, there were further embarrassments on the way. Teller convinced himself and, more importantly, President Dwight Eisenhower that such a thing as a "clean" nuclear bomb could be produced; he probably first got the notion as an argument to use against the imposition of a nuclear test-ban treaty, so that he might carry on detonating his beloved bombs. Teller persuaded Eisenhower and the US public that the development of a "clean" nuke was just around the corner; in fact, such a device has never been created.

With the concept of the "clean" bomb firmly implanted in everyone's minds, the possibility opened up of using nuclear detonations for peaceful purposes. Teller's first project was the creation of a new harbour near Point Hope in the remote north of Alaska. Six "clean" bombs, totalling some 2.4 megatons, would do the job, Teller claimed – and would herald the dawn of a glorious new H-bomb utopia for mankind. Substantial government money – a recurring feature of all Teller's projects – went into the scheme between the late 1950s and 1962, when the Administration of President John F. Kennedy axed it. Typically, Teller had forgotten that the Inuit inhabitants of the Point Hope region might have something to say in the matter.*

---

* A few years later it was shown that in fact Inuit already suffer unduly from radioactive fallout, which is absorbed by the lichens that are eaten by the caribou that the Inuit in turn eat.

Teller then turned his attentions to the notion of controlled nuclear fusion as a source of energy. Again, considerable amounts of US taxpayer dollars were bestowed on Livermore for a project that went nowhere. By the late 1980s Teller himself was prepared to admit that his team's efforts were doomed, yet in 1989, when Pons and Fleischmann announced to a breathless world that they'd achieved cold fusion (see pages 65ff), Teller was one of the first to put his weight behind their claims. He set his Livermore protégé Lowell L. Wood Jr the task of reproducing the Pons–Fleischmann experiments. All Wood succeeded in doing was nearly blowing his laboratory to pieces – not for any fancy-dancy nuclear-fusion reason but because the hydrogen he was using caught fire.

Ever since the split with Los Alamos back in the early 1950s, Teller had lacked the benefit of being surrounded by peers who would evaluate and challenge his ideas. He had made himself largely a pariah in the world of physics by instigating the split from Los Alamos and particularly by his smearing of the much-liked and -respected Oppenheimer. There were difficulties involved in recruiting top-calibre physicists to Livermore; some came nevertheless, but Teller became surrounded more and more by those handpicked to agree with the Great Man. Because of the type of scientist he was, it was essential those working with him should be willing to *dis*agree with him. Hence the numerous failures amid the occasional successes of the Livermore team; hence, too, the untold billions of US Government money wasted on those failures.

By the far the most expensive of his many grandiose ideas was SDI. Teller was probably not the first to conceive that the way out of the dead end represented by Mutual Assured Destruction between the two superpowers was to mount a space-based defence system that would destroy the enemy's missiles before they reached their targets, but he was certainly among the earliest. His initial vision was of a series of satellites armed with X-ray lasers whose beams would zap the oncoming missiles. The X-ray laser had been developed at Livermore in the late 1970s by George F. Chapline Jr, with much of the theoretical work reluctantly done by the pacifist physicist Peter Hagelstein; eventually there were two rival designs.

The beam of an optical laser can do a certain amount of damage, as is well known. Using shorter-wave (more energetic) electromagnetic radiation than light would create a laser capable of far greater destruction, and X-rays had almost the shortest wavelengths of all. (Even better would be gamma rays, a fact that gave rise to dreams of the gamma-ray laser, or graser – see page 128.) The only trouble was that, in order to get an X-ray laser started, you had to use a nuclear explosion. Since the explosion would almost instantaneously annihilate the laser, this was obviously a one-shot-only weapon; it would also, to say the least, be somewhat cumbersome, especially if actually used.* Despite all the difficulties, in November 1980 in an underground Nevada test there were ambiguous indications that both Chapline's and Hagelstein's X-ray lasers might have worked.

Teller and Wood, casting all doubts to the wind, proclaimed the test a triumph: a practicable X-ray laser was, they told their paymasters, just around the corner. Their proselytization did not fall on deaf ears, because the US election of 1980 brought Ronald Wilson Reagan (1911–2004) to the White House. The popular image of Reagan as a genial old duffer conceals such less pleasant aspects of his presidency and politics as his support for some of the vilest dictatorships of the 20th century (including that of Saddam Hussein in Iraq), his financing of South American death squads, his fervent opposition to civil rights legislation, his illegal deals with the Iranian mullahs and Nicaragua's terrorist Contras, his devastation of the poorest people in the US through the axing of liferaft social

---

* These considerations would be less important if the detonation were performed in space, of course; further, a single explosion could be used to trigger a multiplicity of independently targeted lasers. However, you still have to lift a fairly massive – and hazardous! – payload into orbit.

programs, and so on. The image also obscures the fact that Reagan was passionately interested in science and technology. Match this fascination with Reagan's paranoid detestation of Communism and his gullibility (the image was not entirely inaccurate), and there could hardly have been a US President better tailored as the mark for Teller's spiel.* Using an exciting futuristic technical gadget, the X-ray laser, the US could build an impenetrable shield to defend itself against the commie threat. Mutual Assured Destruction would become a thing of the past: since the nukes could never get through, no nation would think to launch them.

Reagan didn't need much convincing, and he put the full weight of his Presidency – and unimaginable amounts of money – behind the development of SDI. Others, however, were considerably more sceptical, both scientists who believed the system could never be got to work and strategists who, knowing nothing of the science, feared such a development could increase rather than reduce the likelihood of nuclear war. The reasoning of the latter was that any nation in possession of SDI could with impunity rain nuclear destruction upon its foes. The obvious course for said foes was to launch an assault upon the nation concerned while SDI was still in the process of development.

Such doubts were discounted. In March 1983 Reagan gave his famous "Star Wars" speech, and out the money began to flow.

In the event, the X-ray laser dropped out of contention fairly early on, and other systems came to be preferred, among them chemical lasers and neutral particle beams. Nonetheless, without Teller's unjustified claims for the X-ray laser it is reasonably certain the program would never have got off the ground. Over time, and particularly after the collapse of the Soviet Union, the program's aspirations were trimmed back, the focus being less on defence against

---

* There were further factors involved. As a primitive Christian, Reagan had since the 1960s taken literally the *Revelations* prophecy of a final Armageddon, which he came to believe referred to a nuclear holocaust. He seems also to have believed around 1980 that the Soviet Union already had an effective antiballistic-missile shield in place, or nearly so.

an all-out nuclear assault, more on protection against an accidental launch. Even here there were difficulties: in 1988, reporting to the House Democratic Caucus, Navy physicist Theodore Postol stated that at best the system might destroy a launch of up to five missiles. Furthermore, any space-based system would be remarkably easy to counter, since the first stage of a pre-programmed assault would be to knock out the satellites bearing the defensive weaponry.*

After the 1991 Gulf War, US attention shifted from SDI to the possibility of ground-based defensive systems. This was because the Patriot inter-ceptor missile had supposedly shown astonishing success in destroying Scud missiles aimed by Iraq at Israel: over 50% of the Patriot's targets were nullified, with initial US estimates set far higher than that. As Richard Cheney (b1941), much later to be US Vice-President but in 1991 George H.W. Bush's Secretary

of Defense, stated: "Patriot missiles have demonstrated the technical efficacy and strategic importance of missile defenses. This underscores the future importance of devel-oping and deploying a system . . . to defend against limited missile attacks, whatever their source." Interestingly, however, Israeli observers, who had the advantage of being able to count the holes in the ground where the Iraqi Scuds

---

* In 2000 Postol similarly took up the cudgels concerning an antiballis-tic-missile sensor, manufactured by the company TRW, which had been declared successful after a 1997 test. In response to the claims of an earlier TRW whistleblower, a 1998 investigation reported that, though the sensor had indeed failed, TRW's software had functioned adequately. When Postol received a censored copy of this curious report in 2000 he rang the alarm bells, claiming possible "research miscon-duct". An investigation into this further claim was delayed by such tactics as classification of materials "for reasons of national security" until time ran out.

had landed, were reporting that the Patriot missiles had at best succeeded in a single interception, with zero being a more likely total. Later the House Government Operations Subcommittee on Legislation and National Security reported, with heavy understatement:

> The Patriot missile system was not the spectacular success in the Persian Gulf War that the American public was led to believe. There is little evidence to prove that the Patriot hit more than a few Scud missiles launched by Iraq during the Gulf War, and there are some doubts about even these engagements. The public and the Congress were misled by definitive statements of success issued by administration and Raytheon representatives during and after the war.

Cheney's deliberate dismissal of contrary experimental results could hardly be more symptomatic of the political corruption of science.

SDI REDUX

The Missile Defense Agency – a modern analogue to the abandoned Star Wars/SDI scheme – is currently spending $10 billion p.a. on such projects as the Space-Based Interceptor, which violates all existing treaties against the militarization of space, and that old favourite of SDI, the space-based laser weapon. To date all of its tests have been embarrassing duds; the Bush Administration's response to this pattern of failure has been to quietly remove the MDA's obligation to report on its progress to Congress . . . while its annual budget is scheduled to rise over the next decade to $18 billion.

As many have pointed out, for about $10 billion the entirety of the world's currently vulnerable depots of fissile material could be secured, thereby almost eliminating the risk of terrorists gaining access to it. There is an existing US program designed to do exactly that, at least for the former USSR: the Cooperative Threat Reduction (CTR) program, inaugurated in 1991 at the end of the Cold War. Since it is regarded as a foreign-aid rather than a defence program, however, it is vulnerable to foolhardy budget-trimming: in 2006 its budget was slashed to below $400 million p.a., its

Eat your heart out, Buck Rogers

lowest level since the program's inauguration (and lower still, of course, because of inflation). It's not hard to see which of the two programs better serves national – and global – safety; it's equally not hard to calculate that the ineffective, unproven program costs some 25 times as much annually as the more effective one. We can only hazard the guess that the reason for the disparity is that boring old CTR lacks the whiz-bang sex-appeal of all that promised shiny new Buck Rogers weaponry . . . or that palms are being greased somewhere.

Meanwhile, despite the existence of various international agreements designed to control and reduce (and eventually eliminate) nuclear weapons, one of the Pentagon's current pet projects is Complex 2030. On the surface this scheme seems innocuous enough: it is, while closing aged plants, to rationalize, update and "consolidate" the US's existing nuclear-arms stockpiles, thereby supposedly making them both safer and a more effective defence tool.* Scratch the surface, though, and oddities begin to emerge, such as that the scheme also includes the opening

---

* Since the stockpiles dwarf those held by all of the rest of the world put together, and are many times more than required to render the entire earth uninhabitable, it's unclear who this defence should be against.

of a new plutonium facility capable of producing over 100 new nukes per year.

HAFNIUM, NO; BANANAS, YES

If the still-ongoing saga of SDI epitomizes the incredibly expensive denial of science, that of the proposed hafnium bomb beggars belief.

In 1998 a physics professor at the University of Texas at Dallas, Carl Collins, bombarded a few specks of the atomic isomer hafnium-178m2 with X-rays from a retired dental X-ray machine in an attempt to trigger an energy release. Atomic isomers (AIs) are analogous to chemical (molecular) isomers: in the latter the atoms in the molecule adopt different spatial configurations than in the "standard" form, whereas in the former it is the particles within the atom that are differently configured. A property possessed by AIs is that they can become "charged up" with energy – large amounts of it – which they release gradually in the form of gamma rays. This made them of considerable interest to scientists following the Grail of the gamma-ray laser, or graser: if an X-ray laser could be almost unimaginably powerful, just imagine a laser that uses the even higher-energy gamma-rays! The problem was how to trigger the AI such that it released its considerable stored energy quickly rather than through gradual decay.

Much money was spent by the US Government on the graser before finally funding was cut; by the time Collins was ready to perform his experiment on Hf-178m2, he was forced to use accounting shenanigans merely in order to obtain a minute sample of the isomer – hence the retired dental X-ray machine, with the sample being placed atop an upturned styrofoam cup.

Despite the constraints, Collins reported success in triggering the Hf-178m2. Even though other experimenters comprehensively failed to confirm his results, the Defense Advanced Research Projects Agency (DARPA) seized on the experiment as grounds for a heavily budgeted pursuit of the hafnium bomb, a bomb that would have an explosive yield comparable to a smallish nuclear fission detonation yet with the advantage of scale: a two-kiloton bomb could be packed

into a hand grenade!* That Hf-178m2's half-life of a mere
31 years would make this the "dirtiest" bomb in human
history was an item glossed over by Collins's enthusiastic
DARPA supporters.

What followed resembles an anxiety dream. Various
investigations, both independent and government-commissioned, cast doubt on Collins's poorly documented and irre-
producible (except by Collins himself and by an old friend
of his, Patrick McDaniel) results and stated categorically
that the physics was bunk. Several of the US's most expert
and trusted scientists went to bat along the same lines. The
response of DARPA was to proclaim to the world that
research into hafnium triggering was going swimmingly, the
only remaining hitch being the need to discover a cheaper
way of producing the phenomenally expensive Hf-178m2.
In one instance, Martin Stickley of DARPA commissioned a
report on Collins's experiments from the Oxford University
physicists Nick and Jirina Stone, expecting it to be a glowing vindication of his claims; in fact it was damning. Rather
than moderate the claims, Stickley swept the report under
the carpet, where it remained until October 2004, when
journalist Sharon Weinberger succeeded in prising it from
DARPA's grip under the Freedom of Information Act.† At
one point even Collins himself tacitly admitted the hafnium
bomb was a non-starter: so far as he could establish, only
about one in 600 of the X-ray photons with which he was
bombarding his sample actually triggered anything, which
meant more energy was required for the triggering than was
being released by it. A bomb is not much use if you have to
use a *bigger* bomb to detonate it.

Matters were further complicated by Collins's and
DARPA's refusal to run a control experiment – for example,
by bombarding a sample of ordinary hafnium with X-rays to
see if the same results emerged, in which case the obvious

* It seems it was some years before it occurred to anyone that a two-kilo-
ton-yield hand grenade is useless except to suicide bombers – ineffective
suicide bombers at that, because, as someone eventually pointed out,
the hard radiation coming off the bomb would kill users before they got
the pin out.
† See Weinberger's book *Imaginary Weapons* (2006) for a definitive
account of the whole hafnium bomb affair.

suspicion would be that the results were a product not of triggering but of something else, such as faulty equipment. Running a control is standard procedure, especially when an experiment yields a controversial result; that those involved refused point-blank to do this was, to say the least, surprising.

The outflow of taxpayer money continued for several years, spurred ever faster by military nonscientist enthusiasts, by US paranoia over the (seemingly nonexistent) researches being pursued by other countries into the hafnium bomb, and by the sensationalist media, until finally reason prevailed: in late 2004, the relevant committees in Congress and the Senate put their collective foot down, and DARPA's funding for research into the hafnium bomb was terminated. Yet a hard core of hafnium-bomb aficionados continue to pursue the dream – or nightmare – and it is suspected covert research is still being sponsored by the Pentagon.

RED MERCURY UNDER THE BED

No one is quite sure where the notion of "red mercury" came from, but it featured in any number of media scare stories from the late 1980s onward and was also – seemingly *is* also – taken seriously by various governments, even though the substance, if it exists at all, is of completely unknown nature. Popular notions are:

- it can be used to facilitate the production of enriched nuclear fuel for weapons;
- it can perhaps be detonated itself as a sort of sub-nuclear but nevertheless very powerful bomb;
- it is a ballotechnic (an explosive chemical) that can be used in place of the fission-bomb trigger for a fusion (hydrogen) bomb;
- "mercury" is merely a codename for fissile material (plutonium, perhaps);
- it is a stimulated nuclear isomer along the lines of hafnium-178m2 and thus a potential explosive source of gamma-rays; or
- it is a paint that, applied to stealth bombers and the like, helps them elude radar.

The list could be extended considerably. Even the name of the putative material is confusing: initially the substance was black, but called "Red mercury" because it supposedly originated from the debris of the old Soviet Union, but soon it lost the capitalization of the "R" and, obligingly, thereafter was red in colour. It could be a solid, a liquid or a powder.

One tempting theory is that the KGB "invented" red mercury as the basis for a sting operation aimed at finding out which rogue states and terrorist organizations were in the market for nuclear materials. A variation was that Russia was engaging in a scam designed to raise billions in foreign exports of a worthless substance, and that it was *Western* intelligence agencies who opportunistically used the sales as a means of finding out who was seeking nuclear materials. A 1995 book, *The Mini-Nuke Conspiracy: How Mandela Inherited a Nuclear Nightmare* by Peter Hounam and Steve McQuillan, claimed South African scientists had created red mercury during the apartheid era, and using it had built countless tactical nukes which were now in the hands of antigovernmental right-wing extremists.

Most certainly there were people in the market for red mercury, with fairly large sums of money changing hands all through the 1990s and even today. In 1997 the *Bulletin of the Atomic Scientists* reported that the buying price ranged between $100,000 and $300,000 a kilo. Earlier, in 1992, Russian President Boris Yeltsin (1931–2007) signed a decree licensing the Yekaterinburg company Promekologiya to produce and sell 84 tonnes of red mercury to the US company Automated Products International of Van Nuys, California. The decree was cancelled a year later and it seems likely no material in fact changed hands; even so, Promekologiya's head reported his company received foreign orders for red mercury totalling over $40 billion. As late as 2004, in a sting mounted in the UK, three men were arrested for attempting to buy a kilo of the stuff for £300,000; although the IAEA made a public statement that red mercury was a hoax, the prosecution insisted, when the case came to trial in 2006, that this didn't affect the men's guilt: they were, however misguidedly, trying to obtain the material for terrorist purposes. (They were acquitted.)

All of this activity for red mercury, yet still nobody knew what it was or what it could do! Well, some people thought they knew. In a paper published in 2003 in *Natsionalnaya Bezopasnost i Geopolitika Rossii*, A.I. Khesin and V.A. Vavilov claimed, according to a summary by the Center for Nonproliferation Studies, "that red mercury can be used to resolve the ills of the human race and planet earth by aiding in oil extraction, restoring exhausted mines to production, reviving unproductive agricultural land, recultivating nuclear test sites, cleansing land polluted with radio-nuclides, producing medicine, and creating environmentally clean fuel for new sources of energy." Wow!

OSMIUM-187

Rather like red mercury but a little later, the isotope osmium-187 became a much-sought item on the international terrorism market as an essential material in the manufacture of nuclear weapons, although its use in said manufacture was somewhat vague. Unlike red mercury, Os-187 certainly exists: it's one of the seven osmium isotopes that exist in nature, although in low concentrations by comparison with the standard form of the metal. It is non-radioactive, but extremely dense – osmium is the heaviest known element – and that density accounts for one of its purported weapons uses: it could be used as the tamper in a bomb, the tamper being the material that inhibits the explosion for as long as possible (not long) in order to increase the scale of the final bang. Unfortunately for such an argument, Os-187 is – at $50,000–$100,000 per gram – ludicrously more expensive than alternative materials that are just as effective for this purpose. Another putative use for the isotope might be as the bomb's neutron reflector, which increases the device's yield, but Os-187 is too dense to make a very effective neutron reflector: the lighter and far cheaper beryllium does a better job.

So why did rogue states and terrorist groups get it into their heads that Os-187 was a must-have item? A clue might be found in the fact that the methods used to extract Os-187 from its parent metal are very like those used to enrich uranium. This similarity seems to have spread the misconception among not only terrorists but also the media and even some (mainly Russian) politicians that, like enriched uranium, Os-187 must have nuclear applications. In 2002 Viktor Ilyukhin, a member of Russia's Security Committee, accused Kazakhstan of unlawfully producing and selling Os-187 for weapons purposes; it seems Kazakhstan itself has concerns about such a potential, because in that nation the substance is controlled.

The scam here is that Os-187 is a completely innocuous isotope: although its presence in natural osmium is a mere 1.64%, that would still be enough to make you nervous of, say, the nib of your fountain pen or the filament of your electric lightbulb were Os-187 substantially radioactive. It has no sensible uses in nuclear weapons. And yet illegal international sales of it continue apace . . .

PSYCHOTRONIC WARFARE

Expenditure by the US Government on research into "psychotronic weapons" – weapons that use psi powers – may seem on the face of it just plain barmy, but things aren't quite as simple as that. Consider this in context. One of the tasks of any government is to try to ensure national security. The existence of psi powers and the like is the longest of long shots; on the other hand, in terms of the overall defence and security budget, the government's investment in psychic research is just the tiniest of drops in the ocean, a minute fraction of a percent. Much of that enormous budget is spent on research into weapons and defences that we *know* won't work, like the hafnium bomb and SDI. If there's the remotest possibility that "there could be something in" psi powers, and especially if there's good reason to believe the enemy is actively carrying out a research program in the field (as there was when the main perceived enemy was the USSR), then wouldn't any government be guilty of gross dereliction of duty if it *didn't* expend some effort in following suit?

That said, bafflement is the only possible response to some of the particulars. For example, one US military project in WWII focused on trying to telepathically influence seagulls to poop on the periscopes of U-boats, thereby obscuring the German submariners' view. Later, in the 1960s, the CIA investigated the possibility of mentally

Russell Targ

controlling cats so that, appropriately equipped with a microphone, the felines could eavesdrop on the enemy's conversations: two spies sitting in a park are unlikely to think twice if an affectionate pussy cat ambles up to them demanding to be stroked, are they? The CIA underestimated cats. When they first tried the system out the cat promptly deserted its post and had a fatal encounter with the traffic on a nearby road.*

The involvement of the CIA in psychic research can be said to have begun in 1972 with a meeting between people from the Office of Scientific Intelligence (OSI) and Russell Targ, a maverick physicist who was co-founder with another such, Harold Puthoff, of the Stanford Research Institute (SRI).† The OSI – like the Defense Intelligence Agency (DIA) – was concerned about reports from the USSR of supposed psychics being investigated for their potential in intelligence gathering. Targ apparently showed them film of people moving objects around tabletops. OSI contacted other departments, including the Technical Services Division (TSD, which had already done some ESP research), to check out if they were prepared to contribute funding to further investigate Targ's claims.

The most fruitful area seemed to be remote viewing, the claimed clairvoyant ability of certain psychics that they can "visit" distant places: spies who could astrally wander around in the enemy's secret bases and report back would

---

* A full account of the craziest US military researches into psychic warfare, including the formation and funding of a unit called the First Earth Battalion, is far outwith the scope of this volume. Readers are referred to Jon Ronson's *The Men Who Stare at Goats* (2004) for an often hilarious, often mind-boggling but ultimately chilling informal survey.
† The Stanford Research Institute is not part of Stanford University.

obviously be an invaluable intelligence asset. Accordingly Puthoff brought to the SRI the New York artist Ingo Swann (b1933), whose psychic abilities were already becoming widely known, and subjected him to a series of tests, in all of which, reported Puthoff, Swann performed spectacularly. Various agencies provided modest funding for further experiments involving Swann and others, notably Pat Price (d1975), a freelance building contractor who lived not far from the SRI.

In the initial remote-viewing experiments, Swann, Price and the rest were told to try to "visit" locales where SRI personnel had been sent as "beacons". It was apparently Swann who pointed out that this approach was, well, a bit useless: if the CIA could plant "beacons" in enemy installations, why bother with remote viewing? Instead, he proposed a technique dubbed "scannate" (scanning by coordinates), in which the remote viewer would be given a set of map coordinates and told to describe what s/he "saw" there. In May 1973 Puthoff was given by an officer of the OSI a set of coordinates that had in turn been given to him by an officer of the CIA: putting this double barrier between the CIA officer and the remote viewer about what lay at the coordinates would seem to obviate cheating. In late May and early June first Swann and then Price were given the coordinates, and they offered quite similar descriptions of what they "saw" there – a military base of some kind – with Price's description being the more detailed. Subsequently Price was asked to "revisit" the site and seek yet further details; he produced an impressive list.

The CIA officer, on being told of all this, laughed: he'd given his colleague at the OSI the coordinates of his vacation cabin in the Blue Ridge Mountains. Puthoff's OSI contact, however, was not content to leave matters there: the similarity between Price's and Swann's descriptions, plus all the extra detail Price had supplied, niggled at him. In due course he found that near the cabin in question was the US Navy communications facility at Sugar Grove, West Virginia, which doubled as a covert National Security Agency (NSA) site. The details of this facility did indeed appear to match what Price in particular had "seen". A report from the OSI to the CIA in October 1973 recounted this, plus the less accurate but nonetheless moderately impressive results of remote-viewing "excursions" to a couple of foreign sites.

The report's author ignored some aspects. The details Price had reported of the Sugar Grove site were largely correct . . . but many were out of date by about a decade, and those that were not were elements that had not changed during that decade. Had Price been fed his information by someone who'd known Sugar Grove but hadn't been there in a while? One possible source was Puthoff himself, who'd worked for the NSA during the early 1960s. This is not to accuse Puthoff of cheating. There's a technique known to conjurers and sham psychics as "cold reading", whereby a skilled interrogator can draw forth from people the most astonishing details through a series of apparently innocuous questions. It's at least possible that Price "cold read" Puthoff – or of course Price may have had another, likewise some-what dated source. Since none of the experiments were performed under proper scientific control it's impossible to tell.

Further funding was found from the Office of Research and Development (ORD) and the Office of Technical Service (OTS; the rechristened TSD) for a new program which, accepting the reality of remote viewing as a given, sought means whereby it could be exploited for intelligence purposes. Almost at once scientists at the ORD began to kick up a fuss about the lack of scientific rigour with which the SRI experiments were done. The clamour spread to the scientific community in general; although in October 1974 Targ and Puthoff published a paper in *Nature* on their remote-viewing work it was accompanied – most unusually for *Nature* – by a qualificatory note from the editors express-ing considerable doubts about the vagueness with which the experimental procedures were described.

By then there had been some personnel shifts in the leadership of the OTS and ORD, the newcomers being significantly more sceptical about the program than those they'd replaced. An experiment done with Price in July 1974, using as target a suspicious, newly discovered (by spy satellite) site in Kazakhstan, was essentially a fiasco: he provided a flood of details, but almost all of them bore no relation to the reality at the site. In his solitary success he correctly "saw" and drew a reasonably accurate representation of a gantry crane that was there, but then he did the same for three other gantry cranes that weren't. As one of the experiment's evaluators observed, with so many details being offered it was hardly surprising that Price, by the laws of probability alone, would get something right. (Oddly, Puthoff and Targ continued for years to claim the experiment was a resounding success, and indeed that it was responsible for the SRI receiving continued government funding.)

A further trial with Price involved "visiting" the code rooms of a pair of Chinese embassies that the CIA had succeeded in bugging. At first his results seemed sensational, describing the general aspects of the code rooms with great accuracy, but when it came to the particulars matters got rapidly vaguer. As later evaluators pointed out, the conditions of the experiment had been so sloppy that in the room with Price there were actually CIA officers who knew the details he was "seeing"; once again the possibility of cold reading seems overwhelmingly likely. Price's untimely death of a heart attack rendered stillborn a new trial, this time to remotely view a Libyan installation. Since Swann had departed some while before, that was more or less the end of the CIA's attempts to exploit remote viewing, and within a couple of years the agency was effectively disowning it.

The DIA and Army Intelligence were not so timorous, however, and their experiments in the field continued until at least the mid-1990s – and may still be continuing. As an example, remote viewers were asked to locate Muammar Gaddafi (b1942) prior to the ill conceived 1986 US bombing raid on Libya. Even later, in 2002, the UK's Ministry of Defence spent £18,000 on a pilot project recruiting supposed remote viewers to participate in the "war on terror"; as one might have predicted non-psychically, the results were apparently "too inconclusive" to take the project further.

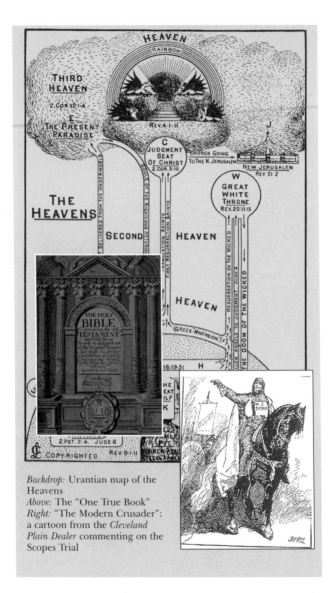

Backdrop: Urantian map of the Heavens
Above: The "One True Book"
Right: "The Modern Crusader": a cartoon from the *Cleveland Plain Dealer* commenting on the Scopes Trial

CHAPTER 4

THE ONE TRUE BOOK

———— ⊰❦⊱ ————

Descended from the apes? Us? How awful! Let us hope that it is not true, but, if it is, let us pray that it will not become generally known!

Emily Sargent (attrib.), wife of Bishop Samuel Wilberforce, on hearing about Darwin's theory of evolution

Most peculiar among the many interpretations of the Bible that I discovered . . . was one that proclaims that the Second Coming of Christ, and the ascent of all Christians into heaven, hinges on the exhaustion of our natural resources. It is a belief that has a complicated relationship to the Bible, for it requires that one believe that God's call to have dominion over the earth is taken literally – but one must also feel that the "End of Times" spoken about in the Book of Revelations is near. Some who concur with such an interpretation believe that global environmental annihilation is a divine requirement for Christ's return.

Stephenie Hendricks, *Divine Destruction* (2005)

IT IS SOBERING to discover that the earth is not called "the earth" at all; its real name is Urantia, as revealed in *The Urantia Bible*, compiled 1928–35 by a splinter group of Seventh Day Adventists from dictation by a committee of alien beings, and first printed in 1955; the text has evidently been edited several times since. In the book's Foreword the aliens explain:

Your world, Urantia, is one of many similar inhabited planets which comprise the local universe of *Nebadon*. This universe, together with similar creations, makes up the superuniverse of *Orvonton*, from whose capital, Uversa, our commission hails. Orvonton is one of the seven evolutionary superuniverses of time and space which circle the never-beginning, never-ending creation of divine perfection – the central universe of

Havona. At the heart of this eternal and central universe is the stationary Isle of Paradise, the geographic center of infinity and the dwelling place of the eternal God.

The seven evolving superuniverses in association with the central and divine universe, we commonly refer to as the *grand universe*; these are the now organized and inhabited creations. They are all a part of the *master universe*, which also embraces the uninhabited but mobilizing universes of outer space.

The religion based on this, originally known as the Urantia Brotherhood but today called the Urantia Book Fellowship, is still extant.

It seems almost a refreshing return to mundanity to move from considerations of Orvonton and its ilk to humble matters like the Divine Creation of our world and ourselves, which will be the primary, although not quite the only, subject of this chapter. A comprehensive survey of the attempted and actual corruption of science by not so much religion *per se* as its more ignorant and/or zealous adherents is far beyond the scope of this book. Instead we shall here look at some examples, focusing largely on Creationism and largely on the US.

THE STATUS OF ANTISCIENCE

At the start of the 21st century, science is in many nations under threat from organized or quasi-organized religion. Most of those nations fall into the category "developing"; the situation is in reality less that science is under threat, more that it has yet to gain a secure foothold. Nations eager to take advantage of the products of science – i.e., technology – still frankly reject large areas of science's essential underpinning, most publicly evolution but often such even more important (at least on a day-to-day basis) disciplines as virology. One assumes those countries will catch up with reality in due course. More alarming, though, is the corrosion of science by religion in the developed nations. Nowhere is the attack more noisy than in the US, where the primary attacker is Christian fundamentalism, the belief that every word of both Old and New Testaments is literally true except the ones you disagree with. Other developed nations face the problem on a minor – but perhaps growing – scale. In the US the problem has become well-nigh institutionalized, thanks in large part, maybe, to the mistaken belief that the ideal of democracy can be applied to issues where it was never intended to be – and *cannot* be – relevant.

Consequently, the status of science in the 21st-century US has reached a nadir incomprehensible in most other developed nations. In March 2006, reporting in the *Broward–Palm Beach New Times* on research by Jesse Bering and David Bjorklund indicating that God is a psychological construct and a product of evolution, Julia Reischel wrote:

> In the 19th century, scientific revelations about the age of the Earth and the development of animal species (and humans) led to the loss of faith of many intellectuals. But the 20th century had a different legacy. While the technological sciences flourished, the end of the century saw science itself increasingly under attack by religious movements, business interests, and, in this country, at least, an antagonistic presidential administration. In a nation where most Americans don't accept evolution at all, science has been under an all-out onslaught.

What's exceptional is that Reischel made the statement not as a matter of controversy but as one of simple fact: "[S]cience has been under an all-out onslaught."

The Creationist broadcaster Ian Taylor, in an undated but recent essay called "The Baconian Method of Science", displays an example of the tortured logic used by the antiscientific in their dismissals of modern science. Antiscientific TV and radio broadcasters might be thought to face a particular dilemma for, if they beat the antiscientific drum too hard, their audiences might start wondering just how the hell it is, if scientists are so stupid, TVs and radios work. The answer seems to be for the broadcaster to claim a fuller understanding of science than scientists have and mount spurious demonstrations of how *science itself* can be used for its own demolition – as if all the world's secular scientists might collectively have been too obtuse to consider points that a lay thinker, guided by God, can come up with by the dozen.*

* This is the same impulse that fuels many a crank theorist: orthodox scientists are too thick to have thought of the possibility that, say, the universe is made of vegetables – a hypothesis proposed by the German cult Vegetaria Universa in the 1960s.

A key point of Francis Bacon's Scientific Method (see page 14) is that the scientist should approach evidence with as few preconceptions as possible; if we truly want to understand how things are, we should first of all strive to clear pre-existing hypotheses from our minds – otherwise all we're likely to find are "facts" that support our own biases. So much would appear to be good sense, and Taylor seemingly accepts it as such . . . but then runs smack into a wall: religious faith is surely one of the biggest of all disqualifying preconceptions. Time for a bit of squirming:

> The major difficulty with the inductive method is that it is an unachievable ideal since man cannot approach a problem with an unprejudiced mind.

This is a reasonable point, and good scientists are aware of it. They must factor into any conclusions they draw that their own conscious or unconscious biases may have played a part. But Taylor goes on:

> The insidious part of "clearing the mind of all preconceptions" is that the good will go with the bad and, if the Bible is the basis for one's worldview, that also will be forfeited. Even if it was possible to clear the mind, the immediate result would be that human reason would flood in like demons to a "house swept clean." The bottom line is that as human beings it is extremely difficult not to have a bias when approaching a problem, so that it becomes a question of which bias is the best bias to be biased with?

Taylor's argument seems to be that the only true way to advance one's knowledge is with a completely open mind, but that this may lead to Godless conclusions. Since Godless conclusions are by definition false ones, the mind should be simultaneously open *and* loaded with religious baggage – i.e., closed.

In another of his undated essays, "The Age of the Earth" (probably written around the cusp of the 20th and 21st centuries), Taylor tackles the matter of the earth's age. He concedes that Archbishop Ussher's figure of 4004BC (see page 152) may not be accurate, but is stalwart in defence of its being approximately correct. His motives are, obviously, to discount the long timescales necessary for modern lifeforms' evolution by natural selection. In particular, he scoffs at science's rejection of the historicity of the Flood; in so doing, he must deny the sciences of stratigraphy and palaeontology in particular, both of which he dismisses as unevidenced.

A few specifics:

(1) Taylor makes much of the erroneous estimate by the geologist Sir Charles Lyell (1797–1875) of the rate of retreat (through erosion) of the Niagara Falls. Lyell, eager to show the earth was far older than Biblical estimates, put the rate of retreat at about 30cm per year, despite contemporary estimates that the rate was more like 60cm annually. Lyell's figure gave him the result that the Falls, to have carved out an 11km gorge, must be about 35,000 years old, putting a 4004BC date for the creation of the earth out of the question. Taylor reports that a more recent measurement of the rate of retreat shows the Falls retreating by about 1.8m annually, giving an age for the Falls of about 6000 years – neatly within Ussher's timescale. What this proves is hard to establish, unless Taylor believes the Falls are necessarily the same age as the earth.

(2) Taylor states quite correctly that comets lose mass each time their orbits bring them close to the sun, and calculates that a body like Halley's Comet loses enough mass at each encounter that it can be no more than a few thousand years old. Periodic comets like Halley's indeed have a limited lifespan; but this has no bearing on any arguments about the age of the universe, since periodic comets are merely those that are knocked by chance gravitational encounters within the Oort Cloud into orbits that bring them relatively close to the sun. To deal with this point, Taylor just flatly denies the existence of the Oort Cloud: "There is not a shred of evidence for it."

(3) The gravitational fields of planetary bodies attract a steady infall of space dust, as one might expect. For this reason, using then-current estimates of the amount of dust scooped up annually by the earth, it was anticipated by many that the early lunar landers would discover the surface of the moon to be covered in a layer of dust perhaps hundreds of metres deep. In the event, the dust on the

moon was found to be mere centimetres deep, and various models had to be revised. In Taylor's view, however, the shallowness of the moon's dust layer is further proof that the moon – and hence the universe – can be only a few thousand years old.

(4) The Big Bang Theory is, according to Taylor, straightforward bunkum: "One obvious difficulty with the theory is the evident order in the universe, galactic walls, precise distances of earth, sun and moon etc. that cannot have arisen from an explosion! Then the great problem of there being insufficient mass in the universe and the need to appeal to 'dark matter' to account for the supposed accretion of the sub-atomic particles."

I am not at all sure what Taylor means by "galactic walls", and it's equally difficult to understand his point about the "precise distances of earth, sun and moon etc.": is he saying that, if the various orbital distances within the solar system were not as they are, the whole lot would fall into the sun? If so, he needs to brush up a little on the relationship between gravitational influence, orbital distance and orbital velocity. Is he saying that the various distances within the solar system are precisely fixed? If so, he's wrong: all are subject to an infinitesimal, but measurable, orbital decay.

In referring to dark matter, in however jumbled a fashion, Taylor does point to what was at his time of writing a genuine cosmological puzzle. It is much less of a puzzle now, just a few years later. The fact that something is as yet not fully understood is no indication that orthodox science is in tatters. Science is all about taking things that are "not fully understood" and working to understand them.

(5) Kelvin, in one of his arguments against Darwinian evolution, said that the earth could not be nearly as old as the theory demanded.* Basing his calculations on the known rate of heat escape from the earth's core, he stated the planet to be no more than 25 million years old. What Kelvin did not know about was radioactivity: it was beyond his (or, at the time, anyone else's) conception that radioac-

---

* He also said that the sun was not old enough – see page 83.

tive decay within the core could be a major source of heat energy. Taylor believes such claims are nonsensical: radioactive decay produces helium, he says, and this "would have diffused through the solid rock filling the earth's atmosphere so that today our atmosphere should be mostly helium with traces of oxygen and nitrogen. Helium would not be lost to outer space." At a guess (and it's only a guess), Taylor has confused radioactive decay (nuclear fission) with nuclear fusion. The most common form of nuclear fusion (as occurs in the sun and stars) is hydrogen-hydrogen fusion, whose product is indeed helium. But the last part of this porridge is perhaps the most puzzling, since Taylor offers no support for his ringing declaration. Why does he think helium wouldn't be lost into space?

(6) The Dead Sea can be shown to be no more than a few thousand years old. Precisely how this affects calculations of the age of the earth is not explained.

(7) Arbitrarily assuming an average rate of 2.4 children per couple, it would take about 5000 years to build up the current population of the world starting with a breeding pool the size of Noah's family. (Elsewhere Taylor states that the Flood occurred some 2000 years after the Creation, so it would seem his calculations have problems other than the obvious.) What Taylor ignores, inter alia, in his calculations is that, for vast swathes of human history, most people born did not reach breeding age.

(8) There are uncertainties in radiometric dating techniques because no one really knows if the decay rates of radioactive atoms are constant over time. It is more likely, says Taylor, that physicists have ascribed very long half-lives to certain isotopes purely in order that extrapolation will give the earth an antiquity measurable in billions of years. Taylor points out that, for example, nearby supernovae can affect the measurements, but ducks the question as to how nearby those supernovae might be . . . a pity because, if under his cosmology the most distant galaxy can be a mere 6000 light years away (otherwise the light from it would take longer to reach us than the 6000-year age of the universe), space must be so exceptionally crowded that many of those exploding stars would be grazing the top of our planet's atmosphere.

(9) Occasionally a lifeform believed to have been extinct for millions of years turns out to be still extant. He's perfectly correct in this. Taylor claims that even the occasional dinosaur turns up, proving that the generally accepted dating of the extinction of the dinosaurs to some 65 million years ago must be false. The fact that some lifeforms can survive for very extended periods of time appears unknown to Taylor. The crocodiles, for example, cheerfully survived the extinction of the dinosaurs and are still around in profusion today.

Overall, Taylor's essay – used here as a type example for countless other modern young-earth texts – is such a mixture of misconception and misinformation that it might seem hardly worth the effort of public dissection, any more than a child's seriously flawed school science essay might be. But this would be to forget that Taylor is not a child in urgent need of private coaching but, as a broadcaster, a figure of some influence. He throws around scientific terminology sufficiently fluently that many of his listeners are likely persuaded by his spurious logic, even though the most superficial of examinations, like the one here, throws up its flaws and inconsistencies.

AN INEVITABLE CONFLICT?

The Judaic tradition, which obviously encompasses also Christianity and Islam, seems always to have had an ambivalent attitude toward the gaining of knowledge, it being ever assumed there are things beyond the bounds of human comprehension: Things Man Is Not Meant To Know. At the same time, it seems accepted that some knowledge that's crucially useful for humankind must be gained *despite* the apparent censure of God: the expulsion of Adam and Eve from Eden after they'd partaken of fruit from the Tree of Knowledge was not without its benefits. Shortly afterwards, Cain murdered Abel and was denied the favour of the Lord . . . yet it was Cain who, according to the Bible, thereafter founded the first city and so gave humankind the valuable gift of civilization, a gift without which the Bible itself could never have been written.

The confusion deepens when it comes to the response

of God to the building of the Tower of Babylon – or Babel. Nowhere can one find a divine prohibition issued against the building of mighty towers, yet God retaliated by inflicting upon humankind the curse of countless mutually incomprehensible languages. The best moral that theological spin doctors can draw from this tale is that the "sin" of the Babylonians was one of presumption: building an edifice which in its mightiness rivalled the works of the Lord. That God is intolerant of human advancement is echoed by many fundamentalists today, although they seem uniformly disinclined to give up their cars, TV sets, microwave ovens and assault rifles. The art is to cherry-pick the bits of science and technology of which God approves or disapproves. The morning-after contraceptive pill is accursed of God, but cluster bombs, landmines and white phosphorus are okay.

A further point here is that, while the Tower of Babel might have been seen in its own age as so impressive as to rival God's own works, today it would be regarded, at least in scale, as trivial. What today might seem like advancements of science and technology into the realm properly reserved for God will likewise, eventually, be seen as humdrum – as basic stuff. It's just that sometimes "eventually" can mean a depressingly long time: it's well over a century since Darwin established the basic mechanism of evolution, and yet many citizens of supposedly developed nations still resist this essential truth on strictly irrational grounds.

In so doing they are, of course, perpetuating an attitude of considerable venerability within the Christian tradition. No lesser authorities than St Paul and St Augustine regarded curiosity as an original sin, the one responsible for Adam and Eve being expelled from the Garden, and they warned against it: the Devil was everywhere, and human curiosity would sooner or later lead the possessor of an inquiring mind into his embrace. The great achievement of Francis Bacon is generally regarded as his enunciation of the Scientific Method, but greater even than that was his breakthrough in expanding hugely the territory that it was accepted as legitimate for human beings to investigate without trespassing into the domain of the Lord. While Bacon still reserved strictly theological matters as being the rightful province of God alone, he proclaimed, and succeeded in convincing his contemporaries, that, not only was it humanity's right to probe the workings of the natural universe, it was divinely approved to do so. God had not put us here to remain ignorant (but worshipful) brutes: the Lord *wanted* us to explore His wonders.

There is another fundamental difficulty that religion – all religion – has with science. It is hard to think of any aspect of science that does not have *time* as an important component; the time may be very short, as in a quantum event, or it may be exceedingly long, far longer than we relatively short-lived creatures can easily cope with: when we consider the creation of the universe, the development of the oceans, the evolution of life or even the formation of the Grand Canyon, then even the longest "human" unit of time measurement, the lifespan, becomes uselessly small. Within science's relationship with time lies the concept of an exceptionally long past and, equally, an exceptionally long future. The Judaic religions (Judaism and Christianity) have at their heart a problem with this in that they are essentially deniers of time: the future will last only until the coming (or second coming) of the Messiah, and thereafter all the rules will change such that the idea of time's passage no longer has any real meaning. The Muslim's dilemma is similar, in that the Messiah has already come, so that everything since – including the future – is a bit irrelevant.* Other religions seek transcendental timelessness, which is again an obviator of one of science's most important foundations. And almost all religions incorporate a similar short-termism when it comes to the past: if a deity or deities created the universe and our planet solely in order to be a home for us, then this surely cannot have happened very long ago – it cannot have been so long ago as not to be realistically measured in human generations.

Perhaps the most damaging consequence of religious short-termism about the future is the attitude of many of the devout that global warming, whatever its causes, simply

* Of course, the vast majority of Jews, Christians and Muslims are intelligent enough to have worked their way around this theological dilemma.

does not matter: there is not enough time left for its full impact to make itself felt. Even were that not the case, we can trust in God to look after us, to avert catastrophe somehow. Similar arguments have been advanced by Christian fundamentalists in various other attacks on environmentalists: God told us to harvest the fruits of the earth, so we should go ahead and destroy the rainforests secure in the knowledge that, since we're following His plan, no harm can ensue. Perhaps these people derive a comfort from their short-termism that's denied the rest of us: they never have to undergo the thought experiment of being confronted by their great-greatgrandchildren, who will have to live with the results of such criminal irresponsibility, because in their worldview the future will be over and done with by then.

The clash between Creationism and science is often portrayed as one between, instead, religion and science. But is this really the case? Such a clash has been by no means universally perceived in the past. Take for example the school of thought called Natural Theology. In high vogue during the 17th and 18th centuries, Natural Theology was the study of the attributes of God as revealed through the study of Nature. It initially arose far earlier, in the writings of medieval scholars like Thomas Aquinas (1225–1274), who essentially spliced Aristotelian and Platonic ideas onto prevailing Christianity in the belief that they were advocating a marriage of reason with faith. To these medieval theologians, the very existence of Creation necessitated there being a Creator (reason bolstered faith), so the study of Nature was really the study of the consequences of the initial designs of what would very much later be called the Intelligent Designer. Similarly, faith-based "knowledge" of the attributes of the Creator would assist in the deciphering of what was discovered in Nature. From this latter interaction arose such concepts as the Chain of Being,* which again dated back to Plato and Aristotle; this notion was

---

* Life exists at every level of complexity from the simplest "animalcules" to the most complex creature of all, the human; beyond humankind lie the angels and God, while below the "animalcules" lie inanimate matter, such as fossils. Each of these levels could be viewed as an essential link of a continuous chain.

important to the medievals, and flowered in the 17th and 18th centuries as an explanation of the world's profusion of lifeforms. Books such as *Wisdom of God in the Creation* (1691) by John Ray (1627–1705) and especially *Natural Theology* (1802) by William Paley (1743–1805) were influential in promoting the concept of Natural Theology; alarmingly, they're still cited approvingly by some Creationists and Intelligent Design proponents today.

Andrew Dickson White

But the myth is deeply entrenched throughout popular culture that science and religion have been mutually antagonistic throughout history. The myth was established above all in 1896 by the publication of Andrew Dickson White's pro-science diatribe *History of the Warfare of Science with Theology in Christendom*. In many ways White was correct: science and religion *should* be at loggerheads, because inevitably improvements in human knowledge reveal that more and more of the basic doctrines of religion are untenable. In fact, however, institutionalized Christianity and scientific progress have for the most part co-existed and even cooperated quite happily. The numerous clashes between the two incompatible forces that White described in his book were almost without exception illusory, many of the tales being products of the two main Christian sects, Catholicism and Protestantism, spinning the facts to make each other look bad.

For example, it's well recognized that by the time Columbus "sailed the ocean blue" in 1492 it was widely accepted in Europe, including by the Catholic Church, that the earth is spherical. Aside from anything else, there are extant European globes that predate Columbus's voyage.

The myth of universal medieval belief in the flat earth seems to have begun with the imaginative biography *The Life and Voyages of Christopher Columbus* (1828) by Washington Irving (1783–1859). Thereafter it was picked up by countless other authors, including White in his *History*.

And it is less than the whole truth to say that Galileo Galilei was persecuted by the Catholic Church for insisting that the earth was not the fixed centre of the universe. The Church was not especially antagonistic to this notion; at the same time, it was not immediately prepared to make a public gesture of embracing the new Copernican cosmology. In short, it was hedging its bets until the picture became a bit clearer: if the supposedly infallible Pope, having had the info from God, proclaimed the earth went around the sun and then later it was proven this was not the case, the Church would be embarrassed. But Galileo, who was a pugnacious type, kept harassing the Church to make a decision. In the end, reluctantly, the Church brought a legal case against him in an effort to shut him up.

Similarly, one of the classic tales within the history of science is that of Giordano Bruno (1548–1600), burnt at the stake for his support of the new Copernican cosmology. Again the tale is less than accurate. In more modern times Bruno would have been regarded as a (probably) harmless lunatic. His cosmology, if such it can sensibly be called, reads like the wildest flights of fantasy, and only *in passim* mentioned that he thought, for reasons immured in his mysticism, the earth went around the sun rather than *vice versa*. This contention of Bruno's owed nothing to Copernicus: indeed, Bruno seems to have despised Copernicus as a mere mathematician. Bruno met his fiery end because of his extreme heresy, which he declined to recant: while of course it's unforgivable that the Church should burn someone to death for disagreement with orthodoxy, the story of Bruno as a martyr in the name of science – with the implicit corollary that the Church condemned scientific progress – is false (although Copernicus's own terror of the Church's treatment of him should he publish his ideas seems to have been quite genuine and not at all unreasonable).

Our perceptions of both the Galileo and Bruno incidents have, then, been moulded more by later propagandists than by the actual historical record. Only a few decades after Galileo's confrontation came an example of the Church in more typical mode. The Danish geologist Niels Stensen (Nicolaus Steno; 1638–1686) announced that studying the rocks and fossils of the earth indicated a far longer history for our planet than anything dreamed of by Archbishop James Ussher (1581–1656), with his estimate of a Creation date of 4004BC based on interpretation of *Genesis*. Far from there being an outcry by the established Church, Steno was promoted up the ecclesiastical ranks, eventually becoming a bishop (and finally, in 1988, being beatified by Pope John Paul II). The vast majority of clerics and the faithful had paid little attention to Ussher's calculation, because for some centuries it had been assumed Genesis was allegorical rather than a literal history.* Even when Biblical fundamentalism began to rear its head in the late 19th and early 20th centuries, there was at first no conflict: one of the early fundamentalist pioneers, William Bell Riley (1861–1947), stated outright that no "intelligent fundamentalist . . . claims that the earth was made six thousand years ago, and the Bible never taught any such thing."

After the initial shock of Darwinism – with clerics like Bishop Samuel Wilberforce (1805–1873) making fools of themselves with their antiscientific protests – the various Christian Churches settled down to a peaceful coexistence with science's new worldview. And why not? It had already been accepted that a *Genesis* "day" was not the 24hr period we're accustomed to – why should God obey the rules of Man? – and the course evolutionary theory painted of the history of life on earth was much the same as that outlined in *Genesis*. By the end of the 19th century, the

---

* Ironically, Steno faced stronger criticism from other scientists than he did from the clergy, since his ideas flew in the face of the established theory of fossils.

few remaining young-earth Creationists might shout a lot but they were regarded as very much on the fringe of mainstream Christian thought.

Andrew White's *History*, despite its multiplicity of false conflicts, laid down a gauntlet on behalf of science, but at first there were not many theologians prepared to pick it up . . . and the truth is that *there still aren't*. While, especially in the US and increasingly in Africa, there are plenty of populist preachers and self-styled Christian demagogues who argue for a young earth and the damnation of Darwinian evolution, theologians *per se* still in the vast majority steer clear of such controversy, seeing in evolution no threat to their faith and in some instances perhaps even a strengthening of it.

The French scientist Pierre Simon de Laplace (1749–1827) presented the Emperor Napoleon (1769–1821) with a copy of the latest volume of his monumental five-volume *Mécanique Céleste* (1799–1825), a treatise on celestial mechanics. After studying it, Napoleon asked Laplace why there was no mention of God in his treatise. "Sir," Laplace is famously reported to have responded, "I have no need of that hypothesis."

The truth is that science does not in fact preclude or discount the existence of the irrational or supernatural: it simply ignores it because it has no need of it. As science progresses, more and more that was once believed to require a supernatural explanation becomes explicable in perfectly rational terms: science is, if you like, progressively taking over the supernatural and turning it into the natural. There are plenty of things science does not yet understand, yet, unlike irrationalism, science does not immediately *identify a specific cause* for the unknown, a cause that is itself unknown: to do so would be no explanation at all.

There were theologians unwilling to compromise. These scriptural fundamentalists tended to belong not so much to the 19th century as to the newly dawning 20th. Perhaps it's something to do with the aftermath of the turning of centuries, times that are especially popular for predictions of the end of the world. Afterwards there must be bitter anticlimax among those who'd confidently anticipated the destruction of all, and a signal that they must

redouble their efforts to counter the appalling sin of
rationalism.

> If one criticism could be leveled against the book, it is that the
> author does not emphasize with sufficient force that the
> Scriptures are so completely our rule of faith, also in the
> matter of creation, that the doctrine of creation does not (and
> ultimately cannot) depend upon scientific evidence or the abil-
> ity to answer scientifically all the notions of scientists, it rests
> on faith alone. The battle between us who believe in the truth
> of God's Word in *Genesis* 1 and 2 and those who have adopted
> some form of evolutionism, particularly theistic evolutionism,
> is a spiritual battle between faith and unbelief, and must never
> be construed as a battle over the scientific evidence supporting
> the one position or the other.

Astonishingly, that passage comes not from the opening
years of the 20th century but from the dawn of the 21st: it
is in fact from a book review by Herman C. Hanko of *Green
Eye of the Storm* (1998) by the Creationist John Rendle-Short;
the review was published in the *Standard Bearer* in February
2001. And Hanko becomes even more focused in his
attempt to explain the extraordinary worldwide influence
Darwin's theory has had on human culture. It is not

> because of the scientific excellence of his theory. It has had to
> be revised more than once. The reason, I suggest, is because
> the theory destroyed the trustworthiness of the Scriptures, and
> especially the very foundation of the gospel in the first chap-
> ters of *Genesis*. And above all because Darwinism abolished the
> need for God and the Christian verities. Thus certainty was
> swept away. Nothing on the earth or in the sky could be guar-
> anteed any more; everything was in a melting pot. Reality was
> nowhere to be found.

These are arguments of a type that perhaps even Bishop
Wilberforce, back when evolution was a brand-new and startling notion, might have thought twice about putting
forward.

The war by religionists against evolution is, in truth,
largely a modern phenomenon, and can be regarded today
as more a political than a religious war – as per the Wedge
(see page 161), the strategic document created by the
Discovery Institute with the intent of using the Intelligent

Design hypothesis as a means of not just subverting the public's comprehension of science but of transforming society as a whole. Nevertheless, while the venom and dirty tricks are to a great extent a product of the last quarter of the 20th century and after, the modern Creationist movement in the US has long antecedents.

THE CRUSADE TO MISEDUCATE THE YOUNG

Where the US Creationists triumphed in the early part of the 20th century was in having the teaching of Darwinian evolution banned in public schools – at least in some states. (In others, many teachers and school boards were intimidated into omitting evolution from the curriculum, a condition that persisted right up until the 1960s, and has recently returned to haunt us.) There have been several battles on this issue, by far the best-known being the celebrated Scopes Monkey Trial of 1925. In fact, this was very much a staged contest. In March 1925 the State of Tennessee had passed the Butler Bill, outlawing the teaching of evolution in Tennessee schools. The American Civil Liberties Union (ACLU), recently formed, saw a chance to live up to its charter. They took out advertisements in Tennessee seeking anyone who'd offer themselves up as a defendant; the town council of Dayton, Tennessee, recognizing the opportunity to gain some publicity, persuaded the science teacher John Scopes (1901–1970) to be the sacrificial goat.

The ACLU hired Clarence Darrow (1857–1938) as Scopes's defence lawyer; he was most famous for having successfully pleaded against the death penalty in the Leopold–Loeb case in 1924. The prosecutor was William Jennings Bryan (1860–1925), a colourful populist politician who had three times run as the Democratic candidate for the US Presidency – in 1896, 1900 and 1908 – being three times roundly defeated. He had tried to push through a bill in Kentucky much like Tennessee's Butler Bill, but had failed.

There is no doubt that Darrow was the brighter bulb of the pair. There is also no doubt that the judge, John Tate Raulston (1868–1956), a devout Baptist, was biased against the defence, for one of his first rulings was that Darrow be

Right: Darrow and Bryan at the Scopes Trial

Left: "He's Always Seeing Things", *Chicago Tribune* cartoon commenting on the trial

not permitted to bring to the stand, as he'd planned, a bevy of distinguished biologists to explain the principles of evolution and how harmless they would be if taught to Tennessee's youth. (Bryan had parallel difficulties: he'd asked a bunch of scientists whom he knew had Creationist sympathies to appear as expert witnesses on his behalf, and they'd all declined.)

Darrow's retaliatory measure for the court's ban was to call Bryan himself to the stand and interrogate him on the principles of Creationism in the – successful – attempt to show them up as nonsense. Bryan made a pathetic exhibition as he stumbled and evaded his way through the examination, presenting views that were not even consistent with themselves. When asked to list geologists with whose earth-history hypotheses he agreed, Bryan could name just two: the maverick amateur geologist George McCready Price (1870–1963), who unfortunately could not be there because lecturing in the UK, and the recently deceased George Frederick Wright (1838–1921), who had written extensively on Creationist geology but whose qualifications were in divinity, not geology.

Bryan's performance would, in short, have convinced few outside Judge Raulston's court. Although Scopes was convicted (as part of the whole staged-trial scenario, he did not have to serve his token sentence), the writing was on the wall for the legal enforcement of Creationist teaching – at least for a few decades.

After the launch of the first artificial satellite in 1957, the Soviet *Sputnik 1*, had offered a dramatic demonstration of the poor quality of science education in the US, the Federal Government became more insistent on, among other things, the proper teaching of evolution in schools: matters had reached such a pass that most school biology textbooks, for fear of cancelled orders worth millions from various populous states where evolution was disliked, either omitted discussion of the subject or blurred it.

As part of its effort in the late 1950s to bring the teaching of biology out of the Middle Ages, a National Science Foundation-funded group called the Biological Sciences Curriculum Study (BSCS) produced a set of textbooks that aimed to bring US school biology up to the level of that of the rest of the developed world. This naturally alarmed Creationists all over the US. Under the direction of Walter E. Lammerts (1904–1996), the Creation Research Society (CRS), one of several influential US Creationist groups, produced a competing school textbook, *Biology: A Search for Order in Complexity* (1970), edited by John N. Moore and Harold Schultz Slusher. After some delay – the text was rejected by all the major US publishing houses – the book was published by the Christian firm of Zondervan. Although there was considerable difficulty in having it accepted in schools (it was overtly religious, therefore its use in a science class would have been in violation of the US Constitution), the book sold very healthily and was undoubtedly influential.

The preface to the CRS textbook was done by Henry Morris (1918–2006), the CRS's president. By the time the book's third edition appeared, in 1974, Morris seems to have come to believe that the way forward for Creationism was to wage a propaganda war, the opening gambit in which must be to face head-on the prevalent (and truthful) charge that Creationism was not scientific but a purely theological fantasy. Changing the fundamental tenets of the belief was out of the question – it is almost a diagnostic characteristic of pseudoscientific "theories" that they are impervious to modification – but a simple semantic trick might do the job. Morris therefore titled a new textbook he edited, aimed at high school teachers, *Scientific Creationism* (1974); the degree to which the change in focus was purely cosmetic can be recognized in the fact that this book was issued in two versions, one from which all theological references had been carefully stripped out, and another, tailored for the more devout, in which they had been preserved. We can view this book as marking the birth of Creation Science, which was not so much a variant of Creationism as a mere retitling of the same old stuff.

Creationism, or Creation Science, as a public concern in the US might almost have died had it not been for Ronald Reagan. While he was still Governor of California, he oversaw a push to encourage the teaching of Creationist ideas in the public schools, and when he began campaigning for the Presidency in 1980s he maintained the theme. It seems certain this was a reflection of his own views rather than merely political opportunism. As President, he appointed a presidential science advisor, George Keyworth (b1939), who was a minor physicist unqualified in the biological sciences but with Creationist leanings, and an education secretary, William Bennett (b1943), who was at least amenable to Creationist notions. Whatever the original motivations, the Republicans discovered that pandering to the scientific ignorance of America's minority (but loud-mouthed) Christian Right was electoral gold-dust. So began that party's war on science, a war that has continued to this day; there is much more about it in Chapter 6(*iii*).

Eugenie C. Scott (b1945) of the National Center for Science Education reported in 1996 on the reactions of a number of teachers. A science teacher in Knoxville, Tennessee, gave this infinitely depressing response: "I'd probably skip the theory of evolution as part of the origin of mankind or the earth . . . We live in the Bible Belt, and it's offensive to some students to hear the theory that man came from monkeys." This despite the fact that a science education which omits "the theory of evolution as part of the origin of mankind" is not a science education at all – presumably the aim of the campaigners.

In 1996 a parent in Lincoln County, West Virginia, complained that the curriculum for a high-school genetics course included the theory of evolution: "Even if it is scientific theory, I don't think it should be taught." Astonishingly, the response of the school board was to scrap the course entirely, and to forbid the teacher concerned from any further teaching of evolution. The explanation for this profoundly anti-educational decision by a body supposedly designed to promote education was offered by one of the teachers involved: "They know a vote for evolution is a vote out of office." It's somewhat unsurprising that in December 2004 Lincoln County's schools superintendent Tom Rinearson was ruefully reporting:

> I'm disappointed in where we are academically overall. In almost every category, our students perform at a lower level than the average across the state. To me that's unacceptable. We need to look below the surface to find out why. Small changes could make a big difference.

They could indeed.

INTELLIGENT DESIGN

The latest variant of Creationism to come to prominence is Intelligent Design (ID), called by its proponents a theory although in fact, technically, at best a hypothesis, without empirical underpinning. To simplify: ID accepts that evolution has occurred, but brings into the equation the notion of "irreducible complexity", which holds that the components of certain organs, bodily structures, etc., would be useless except in combination with the other components, and therefore could not have evolved in isolation – at least not by natural selection: there is no use for the lens of the eye, and certainly no survival advantage in having one,

unless you have the rest of the eye to go with it, and so on.
Therefore, such complex structures, and indeed life itself,
must owe existence to the periodic intervention of an
Intelligent Designer, whom ID proponents are careful for
political reasons not to identify as God. The fallacy of the
basic idea lies in the deliberate ignoring of the fact that
natural selection (a) produces massive redundancies, with
all sorts of useless mutations appearing and in due course
disappearing, and (b) encourages opportunism, whereby
structures that originally served one purpose, or even no
purpose at all, are frequently co-opted into serving another.
In the instance of the eye, the fossil record – and indeed
lifeforms extant today – show all sorts of proto-eyes, which
can be regarded as partway stages along the route to the
modern mammalian eye. The eye that stares up at you from
your grilled trout is an example of a more primitive form of
eye than your own.

ID represents a marginally more sophisticated form of
pseudoscience than Creation Science, and, because its
proponents are sufficiently scientifically literate that they
can utter superficially convincing terminology at appropriate moments, it has persuaded many that it is indeed a valid
scientific hypothesis. In reality it's strongly reminiscent of
one of those off-the-wall quasi-theories science-fiction writers throw into their stories which can be dressed up with just
enough scientific verisimilitude for willing readers to
suspend disbelief, but which no one, including the writer,
expects for one moment to be taken seriously.
Unfortunately, the ID enthusiasts do take their big idea seriously – or at least claim to.

The prime mover behind the ID campaign is the
avowedly right-wing Discovery Institute, founded in 1996.
In connection with the claim by ID proponents that their
hypothesis has everything to do with science and nothing to
do with religion, it's worth examining the logo/banner used
by what was initially called the Discovery Institute's Center
for the Renewal of Science & Culture. The original banner,
introduced in late 1996, showed the classic Michelangelo
image of a bearded God reaching out to touch fingertips
with a rather languid-seeming Adam. As the National
Center for Science Education waspishly points out, "The
image was entirely appropriate, since the Discovery

Institute's president, Bruce Chapman, explained that the Center seeks 'to replace materialistic explanations with the theistic understanding that nature and human beings are created by God.'"

Appropriate it might have been, but there seem to have been concerns about that appropriateness in the context of the claim that ID was science, because in late 1999 the image was altered. Michelangelo's bearded God remained, but in place of Adam we now had a representation of the DNA molecule. This would seem perfectly fitting for an organization intending to put the religion back into science, or whatever, but of course the Discovery Institute is concerned with presenting itself as secular. In late 2001, therefore, Michelangelo was abandoned and a new image introduced, this time a Hubble photo of the Hourglass Nebula, with its eye-like core.

That image is still there in the current logo, but in August 2002 the name of the organization subtly changed: what had been the Center for the Renewal of Science & Culture became the Center for Science & Culture. To quote the National Center for Science Education again: "So far so good. But because the proponents of Intelligent Design have still not published anything in the peer-reviewed scientific literature that supports their claims, there is still a superfluous word in the Center's name: 'Science.' We look forward to the next step in its evolution."

What is most sinister about the whole ID movement is that its intent goes far beyond mere science; it is a political crusade. A strategy document known as the Wedge, compiled by Creationists at the Center for the Renewal of Science and Culture in the late 1990s, spells this out. It reads in part:

FIVE YEAR STRATEGIC PLAN SUMMARY
The social consequences of materialism have been devastating. As symptoms, those consequences are certainly worth treating. However, we are convinced that in order to defeat materialism, we must cut it off at its source. That source is scientific materialism. This is precisely our strategy. If we view the predominant materialistic science as a giant tree, our strategy is intended to function as a "wedge" that, while relatively small, can split the trunk when applied at its weakest points. The very beginning of this strategy, the "thin edge of the

wedge," was Phillip Johnson's critique of Darwinism begun in 1991 in *Darwinism on Trial*, and continued in *Reason in the Balance* and *Defeating Darwinism by Opening Minds*. Michael Behe's highly successful *Darwin's Black Box* followed Johnson's work. We are building on this momentum, broadening the wedge with a positive scientific alternative to materialistic scientific theories, which has come to be called the theory of intelligent design (ID). Design theory promises to reverse the stifling dominance of the materialist worldview, and to replace it with a science consonant with Christian and theistic convictions.

The Wedge strategy can be divided into three distinct but interdependent phases, which are roughly but not strictly chronological. We believe that, with adequate support, we can accomplish many of the objectives of Phases I and II in the next five years (1999-2003), and begin Phase III . . .

> *Phase I*: Research, Writing and Publication
> *Phase II*: Publicity and Opinion-making
> *Phase III*: Cultural Confrontation and Renewal

That scientists and rationalists should object to this plan for the demolition of truth is not surprising; the fact is that many Christian theologians are equally horrified by it, not just from a scientific but from a theological viewpoint. In the *Sydney Morning Herald* for November 15 2005 Dr Neil Ormerod, Professor of Theology at Australian Catholic University, summed this up in an article titled "How Design Supporters Insult God's Intelligence". He describes ID as "an unnecessary hypothesis which should be consigned to the dustbin of scientific and theological history":

> Much depends on what its proponents mean by the term "intelligent design". If they mean that the universe as a whole displays a profound intelligibility through which one might argue philosophically that the existence of God is manifest, their position is very traditional.
>
>   . However, if by intelligent design they mean that God is an explanation for the normal course of events which would otherwise lack scientific explanation, then this is opposed to a traditional Christian understanding of divine transcendence. In seeking to save a place for God within the creation process, the promoters of intelligent design reduce God to the level of what the early theologian Thomas Aquinas would call a "secondary cause".

This is just a more sophisticated version of so-called "creation science", which is poor theology and poor science.

Further bad news for those who argue that ID is a God-blessed hypothesis is that the Vatican doesn't agree with them either – and its officials have frequently said so very publicly. Typical was the article by the Vatican's Chief Astronomer, George Coyne (b1933), in June 2005 in the Catholic magazine *The Tablet*:

> If they respect the results of modern science, and indeed the best of modern biblical research, religious believers must move away from the notion of a dictator God or a designer God, a Newtonian God who made the universe as a watch that ticks along regularly.

At a conference in Florence later that same year he stressed the point more specifically and more robustly, commenting that ID "isn't science, even though it pretends to be". Perhaps even doggeder in his criticisms of "creation science" and ID has been the Austrian Cardinal Christoph Schönborn (b1945). He rightly also has some criticism for the adherents to what he calls "evolutionism", which he defines as a quasi-religious belief system that invokes evolution as a sort of universal cure-all and regards any criticisms of evolutionary theory as "an offence to Darwin's dignity": "The theory of evolution is a scientific theory. What I call evolutionism is an ideological view that says evolution can explain everything in the whole development of the cosmos, from the Big Bang to Beethoven's Ninth Symphony." It's slightly unclear who these "evolutionism" true believers actually *are*, although some scientists might well give that impression as they respond heatedly to the millionth scientifically illiterate attack, but one takes Schönborn's point.

Around the turn of the 21st century the Templeton Foundation, whose aim is to seek reconciliation between science and religion, did some financial sponsoring of conferences and courses to debate ID, then asked the proponents of ID to submit grant proposals for research projects the Foundation might sponsor. But, reported the Foundation's senior Vice President, Charles L. Harper Jr, "They never came in." If even the proponents of ID themselves can't come up with any possible areas of research that might help test their hypothesis, this surely gives the lie to their claims that it is scientific. Or is it simply that they're *uninterested* in experiment and research, accepting the principle of ID as an article of faith without need for such encumbrances as proof? Rather like, say, a religious belief . . .

One criticism of Darwinian evolution often made by ID proponents (and orthodox Creationists) is that, if evolution by natural selection is a reality, why don't we see any evidence of evolutionary changes going on all around us? In fact, there's plenty of such evidence (Darwin's original Galápagos observations showed the effects of evolution over a relatively short period, which was what spurred him to propose his theory in the first place), but every time it's offered to the Creationists they raise the bar: that piece of evidence is somehow *not enough*.

As an example of evolution visibly in action, the "standard" wing colour of various moth and butterfly populations can be observed to change over a very few years in response to environmental factors like soot-laden urban air, the wings darkening in response to the darkening of the places an insect might perch. Such changes can – with difficulty – be explained away within the context of Creationism, but more difficult to discount are researches published from late 2005 onwards in which the effects of continuing human natural selection can be observed directly at the genetic level. A team led by Jonathan Pritchard at the University of Chicago reported in *PLOS-Biology* in March 2006 on their experiments surveying DNA from four different populations: Yoruba (Africa), Japanese and Han Chinese (Asia), and a Utah community (North American, originally of European stock). In the various populations scanned, most genes are universal but a few are still at the stage where they're possessed by some individuals in a population but not by others; the ratio between these two groups gives a measure of how long it has been since the gene arrived on the scene and began to be selected for. Various of the genes chosen for examination concern digestion, and a correlation can be found between their spread through these populations and the shift from hunted/gathered foods to domesticated ones – in other words, to the invention and spread of agriculture among the relevant populations. Similarly, the genes that code for the pale skin colour of Europeans seem to be only about 6600 years old, which means that Europeans retained the dark skin of their African ancestors for some 40,000 years after migrating to the new continent. Again, we're directly observing the fruits of evolution.

A final, more readily observable, everyday example of evolution concerns antibiotics. We're all familiar with the way the doctor, on prescribing a course of antibiotics, stresses we should finish the course rather than abandon it partway once we start feeling better. This is because the full course should kill all the germs that have been making us ill. Stopping before the end can make us very much iller than we were before. Bacteria, which reproduce very quickly, thereby evolve at a prodigious rate. By the time we're halfway through our course of antibiotics, natural selection has already acted to ensure that the remaining bacteria, though perhaps relatively few in number, are considerably hardier and likely more virulent than the original stock. That is, incontestably, Darwinian evolution in action. It is for similar reasons that, in the longer term, use of anti-biotics can lead to the emergence of "superbugs", which have evolved to become resistant entirely to a particular antibiotic or, most dangerously, a whole range of them.

The events leading up to what was hailed as "the new Scopes Trial" – more correctly, *Kitzmiller et al. versus Dover Area School District* – began on October 18 2004 when the members of the Dover (Pennsylvania) School Board voted by 6 to 3 to add the following statement to the area's biology curriculum:

> Students will be made aware of the gaps/problems in Darwin's theory and of other theories of evolution including, but not limited to, intelligent design. Note: Origins of life is not taught.

A month later the Board prescribed that biology teachers must read, at the start of any lesson covering evolution, the following statement:

The Pennsylvania Academic Standards require students to learn about Darwin's theory of evolution and eventually to take a standardized test of which evolution is a part.

Because Darwin's theory is a theory, it continues to be tested as new evidence is discovered. The theory is not a fact. Gaps in the theory exist for which there is no evidence. A theory is defined as a well-tested explanation that unifies a broad range of observations.

Intelligent Design is an explanation of the origin of life that differs from Darwin's view. The reference book, *Of Pandas and People*, is available in the library along with other resources for students who might be interested in gaining an understanding of what Intelligent Design actually involves.

With respect to any theory, students are encouraged to keep an open mind. The school leaves the discussion of the origins of life to individual students and their families. As a standards-driven district, class instruction focuses upon preparing students to achieve proficiency on standards-based assessments.

The *eminence grise* behind the move was a right-wing conservative organization called the Thomas More Law Center, which bills itself as the "Christian answer to the ACLU"; in various areas, from gay marriage to public displays of the Ten Commandments, it has since its foundation in 1999 been picking fights with the ACLU. In order to precipitate a showdown on evolution, the Center had been trawling the US since at least early 2000 for a school board that would be prepared to dictate the teaching of ID in science classrooms, and Dover's swallowed the bait.

On December 14 2004 the ACLU filed suit on behalf of 11 parents, and soon afterwards it was joined by Americans United for Separation of Church and State and by the National Center for Science Education. The latter put out a plea for a legal firm to take on the case *pro bono*, and within hours Eric Rothschild had volunteered the company in which he is a partner, Pepper Hamilton: "I've been waiting for this for fifteen years."

In earlier trials involving attempts to teach Creation Science alongside science, the ACLU's technique had been to demonstrate that Creation Science was not science at all. Obviously ID was vulnerable in the same way. Also, the ACLU team was convinced that the book *Of Pandas and People* (1989) which the Dover School Board had recommended was merely a Creationist text adapted at some early stage to bring it into line with ID; playing a hunch, they subpoenaed the Foundation for Thought and Ethics (FTE), publishers of the book, for any early versions of the text that might exist. This long shot paid off. Most unusually for the publishing world, FTE had kept several earlier versions of the manuscript, whose successive titles alone were revealing: *Creation Biology* (1983), *Biology and Creation* (1986), *Biology and Origins* (1987) and finally *Of Pandas and People* (1987); the final version had been the one published in 1989 (with a revised edition in 1993). Not only were early versions straightforwardly Creationist, much of their content had been transplanted verbatim into the final version with the term "Intelligent Design" being directly substituted for "Creation". These texts were in themselves powerful evidence that, far from being a brand-new scientific hypothesis, ID was merely Creationism under a different name.*

Meanwhile, the legal team for the School Board was having difficulties, primarily because of squabbling between the Discovery Institute and the Thomas More Law Center. In consequence, three of ID's heavyweights, William Dembski, John Angus Campbell and Stephen Meyer, declined to take any part in the proceedings.† The main expert witness left to them was Michael Behe, but in the event his arguments wilted when placed under the spotlight. For example:

Q   Please describe the mechanism that Intelligent Design proposes for how complex biological structures arose.

A   Well, the word "mechanism" can be used in many ways. . . . When I was referring to Intelligent Design, I meant that we can perceive that, in the process by which

---

* The book was written anonymously by Dean Kenyon and Percival Davis. Author Kenyon insists that the book's thrust and motivation are purely towards better science; his co-author, Davis, is a little franker: "Of course my motives were religious. There's no question about it."
† This was an uncanny echo of the difficulty Williams Jennings Bryan had, during the Scopes Trial, in persuading Creationist geologists to appear on his behalf.

a complex biological structure arose, we can infer that intelligence was involved.

Q   What is the mechanism that Intelligent Design proposes?

A   And I wonder, could . . .? Am I permitted to know what I replied to your question the first time?

Q   I don't think I got a reply, so I'm asking you. You've made this claim here: "Intelligent Design theory focuses exclusively on the proposed mechanism of how complex biological structures arose." And I want to know, what is the mechanism that Intelligent Design proposes for how complex biological structures arose?

A   Again, it does not propose a mechanism in the sense of a step-by-step description of how those structures arose. But it can infer that in the mechanism, in the process by which these structures arose, an intelligent cause was involved.

The cases of Justice Cocklecarrot leap to mind.

Behe also got into deep water over his claim that his book *Darwin's Black Box* (1996) had been even more thoroughly peer reviewed than the average scientific-journal paper because its subject matter was so controversial. One of the reviewers he mentioned was Dr Michael Atchison of the University of Pennsylvania. The ACLU's team produced in court an article written by Atchison in which he said his sole involvement had been a phonecall with the book's editor on general matters of content.

Another claim of Behe's was that there had been no serious work done on the evolution of complex biochemical structures like the bacterial flagellum (the example most often used by ID proponents), of the immune system, or of the complicated sequence involved in the clotting of blood. When a stack of peer-reviewed books and papers on exactly these subjects was put in front of him, Behe had to admit that, yes, well, they existed, but he hadn't read them.

His cause was not helped by the fact that his colleagues

at the Department of Biological Sciences, Lehigh
University, while having no argument with his moral or
academic right to embrace the ID hypothesis, at the same
time were less than keen about the publicity it was attracting
to their department, and especially about the way in which
his testimony in the Dover trial, bound to be reported
internationally in the media and to feature in books (like
this one!) for years and probably decades, was likely to forge
an indissoluble link between his name and that of the
university. Accordingly his departmental colleagues took
the unusual step of posting a disclaimer on the department's website:

> The faculty in the Department of Biological Sciences is
> committed to the highest standards of scientific integrity and
> academic function. This commitment carries with it an unwa-
> vering support for academic freedom and the free exchange of
> ideas. It also demands the utmost respect for the scientific
> method, integrity in the conduct of research, and recognition
> that the validity of any scientific model comes only as a result
> of rational hypothesis testing, sound experimentation, and
> findings that can be replicated by others.
>
> The department faculty, then, are unequivocal in their
> support of evolutionary theory, which has its roots in the semi-
> nal work of Charles Darwin, and has been supported by find-
> ings accumulated over 140 years. The sole dissenter from this
> position, Prof. Michael Behe, is a well-known proponent of
> "intelligent design." While we respect Prof. Behe's right to
> express his views, they are his alone and are in no way
> endorsed by the department. It is our collective position that
> intelligent design has no basis in science, has not been tested
> experimentally, and should not be regarded as scientific.

The difficulties of their star witness were not the defence's
only worry. One of the linchpins of their argument was that
this wasn't a dispute between science and religion but
between two different schools of science, and to this end
former Dover School Board president Alan Bonsell and the
chairman of the Board's curriculum committee, William
Buckingham, staunchly denied having any religious motive.
Unfortunately, there was eye-witness testimony of them
both talking about how they wanted to see evolution
"balanced" with Creationism in the curriculum, and even

video of Buckingham saying precisely this on a local TV station. The local newspapers had carried reports of a meeting back in June 2004 when Buckingham had said of the evolution/Creationism "balance": "Two thousand years ago, someone died on a cross. Can't someone take a stand for him?"

The final decision, when it came on December 20 2005, represented a triumph for rationality. US District Judge John E. Jones III issued a 139-page judgement that effectively demolished ID as a purported science. He was blunt about the dishonesty of the pretence whereby a religiously based scheme was being dressed up as something scientific: "We find that the secular purposes claimed by the Board amount to a pretext for the Board's real purpose, which was to promote religion in the public school classroom. . . . It is ironic that several of these individuals, who so staunchly and proudly touted their religious convictions in public, would time and again lie to cover their tracks and disguise the real purpose behind the ID Policy."

He added some harsh words for the tactics of the ID movement in general, words which naturally fell on deaf ears, while also making a pre-emptive strike against the inevitable smear that he would be cast as an "activist judge", a ploy widely used by US right-wing extremists when the law upsets them:

> Those who disagree with our holding will likely mark it as the product of an activist judge. If so, they will have erred as this is manifestly not an activist court. Rather, this case came to us as the result of the activism of an ill-informed faction on a School Board, aided by a national public interest law firm eager to find a constitutional test case on intelligent design, who in combination drove the Board to adopt an imprudent and ultimately unconstitutional policy. The breathtaking inanity of the Board's decision is evident when considered against the factual backdrop, which has now been fully revealed through this trial. The students, parents, and teachers of the Dover Area School District deserved better than to be dragged into this legal maelstrom, with its resulting utter waste of monetary and personal resources.

Truth in science is not something that can or should be

decided in a court of law: reality does not obey court orders any more than it does democratic votes. At the same time, it cannot be denied that the Dover trial did much more than settle a particular legal case: the doctrine of Intelligent Design was effectively shredded, and very publicly so.

No Jobs for the Boys?

A frequent plaint of Creation Scientists is that they are discriminated against by the powers-that-be of orthodox science: their papers are rejected by the scientific journals, or they are even dismissed from their jobs because of their beliefs. These are serious charges.

In fact, there are plenty of examples of Creation Scientists publishing papers in the scientific journals, so their beliefs alone are clearly not grounds for discrimination. Where Creationists have failed to gain publication in the journals is *when they have been advocating Creationism*, and there the "discrimination" is on the grounds of bad science – the same "discrimination" whereby countless other papers are rejected. As a single example of a Creationist widely published in the journals there's Robert V. Gentry (b1933), who did important work on the radioactive "haloes" found in primordial igneous rocks. These microscopically sized rings are produced by radioactive decay during the rock's formation; Gentry convinced himself that they offered proof that the delay between God's creation of the chemical elements and His creation of the rocks could be at most a matter of minutes. Gentry's research was published in several eminent science journals, although the editors of those journals naturally balked at his theological inferences, which had no place in a scientific paper.

In 1982 Arkansas's "balanced-treatment" law – whereby schools were forced to offer both evolution and Creation Science side-by-side – came up before federal judge William R. Overton (1939–1987), and during the trial the Creationists were challenged on their claim that the scientific journals discriminated against them. Asked to produce examples of papers that had been rejected solely because of their writers' Creationist beliefs, they failed to produce even one.

Cases of unfair dismissal are similarly hard to find. One classic example often cited is that of Forrest Mims, who in 1988 was reportedly fired by the magazine *Scientific American* on the grounds of his Creationism. This sounds like discrimination pure and simple . . . except that the story isn't true. Mims was not an employee of *Scientific American* but a freelancer who'd sold articles to the magazine's "The Amateur Scientist" column. There was talk of him taking over the column full-time, but that failed to materialize. As to whether or not his Creationist views played a part in his rejection, it's hard now to tell; Mims – today a very widely published author in the field of electronics – certainly believed this to be so. Whatever the case, the fact remains that he was not employed in the first place, so could not have been fired.

Then there's the case of Jerry Bergman, who in 1979 was denied tenure at Bowling Green State University, Ohio, supposedly because of Creationist beliefs. Well, perhaps. Bergman also wrote a published letter to the newsletter of the racist National Association for the Advancement of White People that the reason for the denial was affirmative action/reverse discrimination. (To hammer home the message, he cited two further examples where he believed he had been the victim of affirmative action, one relating to colour and the other to gender.) Leaving aside the oddity of the letter's place of publication – Bergman is vehement that writing to a racist publication concerning a racially sensitive accusation does not necessarily imply racism in the writer – it's clear that he himself, despite his own later claims, at the very least regarded his denial of tenure, at the time, as being for other than religious/Creationist reasons. Bergman went on to write a whole book, *The Criterion: Religious Discrimination in America* (1984), about people whom he claimed had suffered unfair discrimination or persecution because of their Creationism.

One such was Ervil D. Clark (1927–1981), the son of the pioneering Creationist Harold W. Clark (1891–1986). He failed twice to gain a PhD at Stanford University, both times through a split committee decision over his oral defence of his Masters thesis. While the Creationist literature regards this as evidence of discrimination, Clark himself in later life was far less sure, apparently agreeing at least in part with the committee's strictures concerning his narrowness of knowledge in his chosen field, ecology. Later he gained a PhD without mishap from Oregon State University.

But by far the best known Creationist to have claimed academic discrimination against him is Clifford Burdick (1894–1992), famed for his false interpretation of the dinosaur and "human" fossil footprints at Paluxy River.[*] He served as a geologist with the Creation Research Society and later the Institute for Creation Research. Missing from the "official" story is that Burdick's Creationist colleagues likewise expressed significant misgivings about his work.

In 1966 Burdick announced he had found modern conifer pollens in strata at the Grand Canyon that had been dated to the Precambrian; clearly such an anachronism would cast doubt over the whole basis of geological dating. Even Walter Lammerts, the Creation Research Society's director, had qualms about the report. In 1969, consequently, the CRS deputed two independent scientists to go to the Grand Canyon with Burdick to check his results; they concluded that the anomalous findings were a result of Burdick's incompetence at taking soil samples. This was presented by Burdick to the CRS as his discovery having been confirmed!

Earlier, in 1960, Burdick had claimed to the CRS that the University of Arizona had refused him a PhD solely on the grounds of his Creationism. The truth of the matter emerged during and after the lawsuit which he brought and lost against the university for religious discrimination. Although he had studied at the university, his was not a degree course; even so, he did sufficient work of a suitable calibre to be permitted to sit a comprehensive examination that would, if he passed it, bring him that coveted PhD. All this time he had been keeping his Creationist views hidden from his professors, but a few days before the oral exam one of them mentioned having come across a Creationist article

* For an account of this fiasco, see for example my book *Discarded Science* (2006); a far more exhaustive discussion can be found in Ronald L. Numbers's *The Creationists* (1992).

Burdick had written. According to Burdick, this threw him into such a state of indigestion and insomnia that, on orals day, he was too exhausted and mentally confused to perform adequately. He failed the exam. He also failed in the defence of his Masters thesis at the university, this time because, he said, he had reacted adversely to the medicines he was taking for hepatitis acquired on a field trip.

Lammerts initially intervened on Burdick's behalf with the university; however, when it was explained to him what had actually happened, he backed off hastily. Burdick did go on to make a second attempt to gain a PhD from the University of Arizona, but this time (according to Burdick) his course was disrupted by the discovery by one of the professors that Burdick was the source of the Paluxy River footprints story. Burdick lied his way out of that dilemma, but felt the writing was on the wall for him at the university. In 1966 he gained a PhD in geology from an outfit called the University of Physical Science, based in Phoenix, Arizona. This proved to be not a university at all, just a sort of geological club. Lammerts suggested Burdick henceforth use the rather bizarre "Hon. PhD" rather than "PhD" after his name – even though various other members of the CRS quite blithely listed their "honorary degrees" without any such caveat. The CRS continued to make use of Burdick's services, but cautiously; in the light of his history it's alarming they kept him on at all.

Among the important figures in Creationism who have gained degrees despite their open fundamentalism have been Henry Morris, Lammerts himself and Duane Gish (b1921), to name but three. More recently, in 1989 the openly Creationist Kurt Wise gained his PhD at Harvard despite the fact that he studied under no less a personage than Stephen Jay Gould (1941–2002), the *bête noire* of Creationists everywhere!

The complaints about job discrimination are part and parcel of a more general claim by Creationists and fundamentalists that US society as a whole persecutes them. It's arguably cruel to pick on a single example from so many, but a particularly loopy piece of statistical reasoning was expressed by the US syndicated religious pundit Cal

Thomas (b1942) at the end of 2006. He asked: "Why is it 'in vogue' to disbelieve in a Creator of the universe, who loves us and wants to have a relationship with us, and not 'in vogue' to believe?" On this ground he bewailed the fact that so many Americans espouse atheism – and, hilariously, declared with wide-eyed wonder that it's really quite possible to engage civilly with atheists so long as you know a few tricks of the trade!

How many people would have to believe in God for Thomas to start regarding belief as being "in vogue"? At the time of his writing, polls showed that approximately 90% of people in the US believed in God, with agnostics and atheists together comprising a mere 10%. One cannot credit that he was ignorant of these poll results, which were publicly available and fairly consistent with previous polls over decades. The only possibilities are that either (a) he was incapable of understanding the statistics they represented, or (b) he was committing a "rhetorical truth".

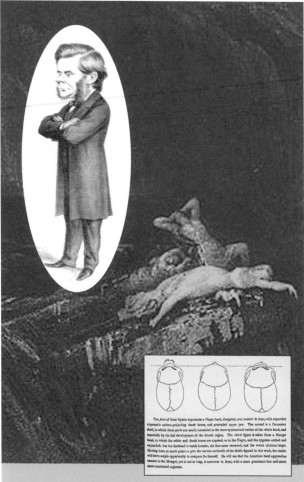

The *first* of these figures represents a Negro head, elongated, and narrow in front, with expanded zygomatic arches, projecting cheek bones, and protruded upper jaw. The *second* is a Caucasian skull, in which these parts are nearly concealed in the more symmetrical outline of the whole head, and especially by the full development of the frontal region. The *third* figure is taken from a Mongol head, in which the orbits and cheek bones are expand, as in the Negro, and the zygome arched and expanded; but the forehead is much broader, the face more recessed, and the whole cranium larger. Having been at much pains to give the *norma verticalis* of the skulls figured in this work, the reader will have ample opportunity to compare for himself. He will see that the American head approaches nearest to the Mongol, yet is not so long, is narrower in front, with a more prominent face and much more contracted zygome.

Top: Caricature in *Vanity Fair* of T.H. Huxley
Backdrop: Detail from *The Flood* by John Martin
Right: Figure by Samuel George Morton showing Black,
White and East Asian skulls

CHAPTER 5

IDEOLOGY TRUMPS SCIENCE

————————— ⟨◈⟩ —————————

The theory is widely accepted within the scientific community despite a lack of any conclusive evidence . . . It should be noted that these scientists are largely motivated by a need for grant money in their fields. Therefore, their work can not be considered unbiased. Also, these scientists are mostly liberal athiests [sic], untroubled by the hubris that man can destroy the Earth which God gave him.

"Global Warming", *Conservapedia*
(www.conservapedia.com), 2007

PERHAPS THE MOST dramatic example of the ideological corruption of science is book-burning, the deliberate destruction of knowledge for political or religious reasons. It is, of course, a form of genocide, in that the lives of all future human generations are maimed by the loss of the knowledge.

In AD645 the third Caliph, Uthman (*c*580–656), ordered the destruction of the Library at Alexandria on the grounds that either the books agreed with the Quran, in which case they were redundant, or they disagreed with it, in which case they were worthless or evil. In an act that has rightly been vilified throughout the centuries since, the Library's books were used as fuel to heat Alexandria's public baths. There were so many books that the burning took six months.

What is less known is that Uthman was far from the first to burn the Library's books, and the crime far from exclusive to Islam. The first three times the Library felt the flames were through acts of criminal negligence: Ptolemy VIII (d116BC), Julius Caesar (*c*100–44BC) and the Emperor Aurelian (*c*215–275) simply didn't care – in 88BC, 47BC and AD273 respectively – that their burning of large parts of Alexandria would destroy the Library along with the rest. After each desecration the Library was reconstituted to a greater or lesser degree. By AD391, a little over a century after Aurelian's crime, there were some 40,000 scrolls, housed in a temple to Serapis. All of these were destroyed in another act of religious madness, this time by a Christian, Archbishop Theophilus (d412), on the grounds that anything contained in a pagan temple must be by definition diabolical.

It was to a great extent because of the loss of the Library at Alexandria that Europe suffered a Dark Age for the best part of the next millennium. The destruction still affects us today: imagine what our world would be like had there not been that long hiatus – and imagine how much misery and ignorance our ancestors could have been spared.

Indeed, mass book-burning – which in more recent times has come to include mass burnings of CDs and the like, as in the wake of the Dixie Chicks' public denunciation of President George W. Bush – have been throughout history almost a commonplace among regimes or mobs whose ideology tells them they are in the exclusive possession of rectitude . . . or, perhaps, who become nervous at the threat presented by reality, as enshrined in books, to such ideological certainty. The Nazis' declared reason for their mass book-burnings in the 1930s was that, since the Reich was going to last for at least 1000 years, it did not need intellectual pollution by the "falsehoods" of such people as Proust, Wells, Freud and Einstein. In truth, of course, the burning was defensive: the ideas of these and many, many other writers whose works met the same fate could far too easily prick the frail bubble of the absurdity that was Nazi ideology. Mao Tse-tung (1893–1976) and Josef Stalin (1879–1953) attempted similar exercises, but with less success. Book-burning's digital equivalent, though less dramatic and less demonstrative of repudiation, is now also practised. At the time of writing, the Bush Administration in the US, frightened by the conflict between its ideology and environmental realities, has recently started a rapid program of making the archives/libraries of the Environmental Protection Agency unavailable, first to the general public and very soon thereafter even to the agency's own scientists. Efforts are underway to stop the process, already begun, of destroying those libraries entirely. Once again, this act in defence of an ideology is being presented in the guise of something being done because of the strength of that ideology.

But it was because of confidence in an ideology that the other great book-destruction of ancient times – the Asian equivalent to Europe's loss of the Alexandria Library – was carried out. In 213BC the Emperor Shih Huang-ti (c259–210BC), the founder of the Ch'in Dynasty, ordered all the books in China to be destroyed because his dynasty, which was set to last 10,000 generations and unify all the world – the claims for the Third Reich seem modest by comparison – would have no need of the irritating disruption that contemplation of the past might cause. Exceptions were made for books on agriculture (which were practically useful), medicine (likewise) and fortune-telling (because the Emperor was really keen on fortune-telling). All the Chinese literature we possess from before 213BC is, accordingly, essentially *ersatz*: copies done by much later writers, relying on memory.* In the event the Ch'in Dynasty endured for rather less than 10,000 generations: just 15 years.

The New World was not spared. Precisely four books survive from the Mayan Empire: all the rest were put to the torch in the 16th century by Christian zealots, led by the Franciscan Diego de Landa (1524–1579), who crossed the Atlantic in the wake of the conquistadors. The aim was to destroy the Mayan culture entirely and permanently, and in essence that aim was achieved. Scholars today know, because the virtuous monks kept self-congratulatory records of the destruction they wrought, that the Mayans wrote down their culture in thousands of books; but our knowledge of the culture itself is sketchy, based on tortuous deduction from desperately scant evidence – writings preserved on pottery and walls. A similar concerted attempt was made by the Serbs in Sarajevo in 1992 to extirpate Bosnian culture by destroying the National and University Library.

* We can see where the scenario of Ray Bradbury's novel *Fahrenheit 451* (1953) came from! It's interesting that this novel has become one of the best-known in the world – to the point where the fallacy that book paper burns at 451°F is routinely repeated – even though, as a novel, it limps. So powerful has the spectre of book-burning become to us that the book's premise alone has assured it a place in our cultural iconography.

It's almost impossible to guess toward which book the attentions of the book-burners and -banners might turn next. It's well known that the innocuous Harry Potter children's novels have drawn the hostility of Christian fundamentalists numerous times because of their "glorification of witchcraft"; plenty of other completely harmless children's fantasies have been similarly pilloried, albeit with less publicity. Still, one can predict that a book with wizards in it is likely to generate heat under some bigot's collar somewhere; but who would have predicted *The Diary of Anne Frank* (1952) might have a similar effect? Yet there have been at least three local attempts in the modern US to ban this book, once on the grounds of sexual offensiveness, once on the grounds that it was pornography, and once, most ludicrously of all, seemingly for no other reason than that it was a "real downer". All three attempted bannings were small-scale and specific, but that doesn't mean the next one will be. And the book to be banned could be your kid's biology textbook, on the grounds that it treats evolution.

What do I mean, "could be"?

The destruction of knowledge is of course the very antithesis of science, whose aim is to increase it. This is true even if the knowledge threatened with destruction is demonstrably false. Without a comprehension of false knowledge we cannot properly appreciate what we have so far uncovered of the truth. The fallacies of past misunderstanding educate us to become better detectives in tracking down possible fallacies in our own worldview.

RACIST PSEUDOSCIENCE

Elijah Muhammad (1897–1975), leader of the US movement Nation of Islam from 1934, taught – claiming the teaching came from his predecessor Wallace Fard Muhammad (1891–*c*1934) – that Black people were the original inhabitants of the world, with Whites a race of troublemakers created later by a mad scientist called Yakub,* working for 600 years about 6000 years ago on the island of

* Since Elijah Muhammad's death, Nation of Islam followers have generally come to regard the tale of Yakub as allegorical, like that of Adam and Eve.

Patmos. These "grafted" people – i.e., Whites – would rule the earth until the Blacks regained their rightful position, which process had begun in 1914. The tale of Yakub is of course scientific nonsense, as is the racism of the Nation of Islam – although at least we can see where Elijah Muhammad's anti-White racism came from: he reportedly witnessed three lynchings of Blacks in his Georgia home-land before he was out of his teens, enough to make anyone believe Whites were degenerate brutes.

"Every part of the body of a non-Muslim individual is impure, even the hair on his hand and his body hair, his nails, and all the secretions of his body." Thus the Ayatollah Khomeini (c1900–1989). Again: "This is why Islam has put so many people to death: to safeguard the interests of the Muslim community. Islam has obliterated many tribes because they were sources of corruption and harmful to the welfare of Muslims."

The notion of racial purity, linked often with religious issues, has ever brought out the worst in humanity: the pages of history are littered with the corpses of those slaughtered simply because they were of the "wrong" race. The pattern is simple, and oft-repeated: first you declare an identifiable minority of your society's population inferior; then, in times of trouble – especially of economic recession – you announce the problems are the responsibility of this minority; then you start the pogroms. The theory's insult to humanity is its refusal to recognize individuals as individuals: they are "blacks", or "Jews", or "gays", or "gypsies", or "Catholics", or "women" – members of categories, not people in their own right.

Some of the pronouncements of the racial purists are truly risible, or would be if they weren't so repellent – like those of the Holocaust deniers. Around the end of the 19th century Charles Carroll (b1849) – who was far from alone in his views – produced three books, *Negro Not the Son of Ham, or Man Not a Species Divisible into Races* (1898), *Negro a Beast, or In the Image of God* (1900) and *Tempter of Eve, or The Criminality of Man's Social, Political, and Religious Equality with the Negro, and the Amalgamation to Which These Crimes Inevitably Lead* (1902), which sought to show that God had created the Black as merely a higher animal whose sole function was to act as a servant to the White. When *Genesis* refers to the tempting serpent it is *really* referring metaphorically to Eve's coloured maidservant. Ellen Gould White (1827–1915), writing in 1864 in fundamentalist vein, sought to answer the question: Why are there diverse races of Man? Only one "type" of human being could have been saved from the Flood – because, of course, Noah's family could not have been made up of more than one racial stock (even if Noah were black and Mrs Noah white, the children would all have shared the same genes). Since the Flood, however, people have committed the unforgivable sin of mating with animals; all races except the aptly named Mrs White's are the results of such miscegenation.

Prewar German theorists produced some of the most bizarre "scientific" racist ideas. The attack was two-pronged – not only the attempt to prove the inferiority of, say, the Jews, but also the more positive struggle to prove the superiority, bordering on perfection, of themselves. Hans Günther (1891–1968), a leading Nazi anthropologist, gave us a few illuminating clues as to the natural superiority of the Nordic type. For example, Nordics wash themselves and brush their hair more often, are better athletes and, if female, keep their legs rigidly locked together when travelling by bus. In the mid-1930s Julius Streicher (1885–1946) proposed that Jewish blood-corpuscles are totally different from those of the rest of us. Why have biologists failed to notice this? Because so many biologists are Jews, and of course they're operating a coverup.*

The racial-purity theory is, in its details, both untenable and dangerous; but what of the broader brush? In fact Mrs White unwittingly pointed up its fundamental fallacy. We all came from common stock; ever since, breeding between diverse human populations has been commonplace. There is no such person as a genetically "pure" individual: we are all mongrels.

The French physician Julien-Joseph Virey (1775–1846), who first put forward the notion of biological clocks and pioneered the study of what is now called chronobiology, had some curious ideas when it came to other subjects.

---

* For more on Nazi racism, see pages 239ff.

Blithely unaware that the aesthetics of the human form might vary from one culture to the next, he maintained that all the ugly peoples of the world were "more or less barbarians" while beauty was "the inseparable companion of the most civilized nations", and, despite (one assumes) complete ignorance of his subject matter, confidently proclaimed that Black women possessed a greater degree of lasciviousness than their White counterparts, a difference he attributed to the Black women's greater voluptuousness. As for intellectual differences between the races, he produced this piece of dazzling illogic:

> Among us [i.e., Europeans] the forehead is pushed forward, the mouth is pulled back as if we were destined to think rather than eat; the Negro has a shortened forehead and a mouth that is pushed forward as if he were designed to eat instead of to think.

Similarly Georges Cuvier (1769–1832), one of the fathers of geology and palaeontology, mapped his own racist preconceptions onto his studies of comparative anatomy. In his *Elementary Survey of the Natural History of Animals* (1798) he wrote that the "White race, with oval face, straight hair and nose, to which the civilized peoples of Europe belong and which appear to us the most beautiful of all, is also superior to others by its genius, courage and activity",  adding that there was a "cruel law which seems to have condemned to an eternal inferiority the races of depressed and compressed skulls". In the phrase "which appear to us the most beautiful of all" he might seem, rare among his contemporaries, to have at least a foggy awareness of the role of cultural relativism in determining our ideas of human beauty, but he crushes this assessment with his further remark that "experience seems to confirm the theory that there is a relationship between the perfection of the spirit and the beauty of the face" – a remark that implies he knew nothing whatsoever about his fellow human beings.

In his *Recherches sur les Ossements Fossiles de Quadrupèdes* ("Investigations of the Fossil Bones of Quadrupeds", 1812) he declared that Africans were "the most degraded of human races, whose form approaches that of the beast and whose intelligence is nowhere great enough to arrive at regular government".

If the Europeans were bad, what about their descendants in the New World? The degree to which racism was endemic among US Whites in the 19th century is hard for us to comprehend today, blinded as we are by the myth that the North fought to free the Southern Blacks from slavery. In fact, at the time it was widely held that Abraham Lincoln (1809–1865) brought Emancipation into the picture solely in the hopes of inspiring the Southern Blacks at last to rise up against the slave owners; and Lincoln's stated ideal was that, after Emancipation, the US Blacks would be "repatriated" *en masse* to Africa. In debate against Stephen Douglas (1813–1861) in 1858 Lincoln made his position clear:

> I will say, then, that I am not nor ever have been in favour of bringing about in any way the social and political equality of the black and white races – that I am not, nor ever have been, in favour of making voters or jurors of negroes, nor of qualifying them to hold office, nor to intermarry with white people; and I will say in addition to this that there is a physical difference between the white and black races which will ever forbid the two races living together on terms of social and political equality. And inasmuch as they cannot so live, while they do remain together, there must be the position of superior and inferior, and I, as much as any other man, am in favour of having the superior position assigned to the white race.

Even after signing the Proclamation of Emancipation in 1863, Lincoln felt compelled to say that "I can conceive of no greater calamity than the assimilation of the Negro into our social and political life as our equal". Remember that Lincoln was, for the society in which he lived, one of the most enlightened of men.

Of course, there's no reason why Creationism should foster racism, but nonetheless many people have observed that the two often seem to go hand in hand. A central cause is the belief that, after the Flood, the earth was repopulated by the descendants of Noah's three sons Japheth, Ham and Shem. It has proven far too tempting to many Creationists to connect these three lineages with the races of mankind. A general Creationist concurrence today appears to be that the lineage Ham sired is represented by the dark-skinned races, although the definition of "dark-skinned" varies a little, sometimes including the Asiatics and sometimes not. Thus we have Henry Morris writing: ". . . all of the earth's 'colored' races – yellow, red, brown, and black – essentially the Afro-Asian group of peoples, including the American Indians – are possibly Hamitic in origin." This would seem harmless enough, if anthropologically unsound, except that we can note in passing that the specious correlation between Blacks and Hamites was used in the previous century as a justification for slavery. Further, as soon as the Creationist mind fastens upon such a hypothesis, it is driven to start extrapolating from it, and therein lies a danger. Here's Morris again:

> Often the Hamites, especially the Negroes, have become actual personal servants or even slaves to the others. Possessed of a racial character concerned mainly with mundane matters, they have eventually been displaced by the intellectual and philosophical acumen of the Japhethites and the religious zeal of the Semites.

In other words, the Hamites are the intellectual inferior of the other two races. To repeat, the fault here lies not in Creationism *per se*, which has plenty of faults of its own, but in the minds of the human beings who are its proponents – just as there are those who attempt to use their research in evolution to promote their own racist ideas. But it seems especially obvious in the instance of Creationism because, as numerous surveys have shown, there is in modern times a correlation between low educational level and Creationism and another between low educational level and racism. Put the two correlations together and the reasoning becomes obvious. In past eras, of course, the highly educated and the ill educated believed in the Creation alike.

Even without invoking the lineages of Noah's three sons, it was perfectly possible to bend fundamentalist views into supporting racist ideas. Considerably before Morris and his ilk there was the prevalent notion (monogenesis) that all human beings were descended from Adam and Eve: on expulsion from Eden humans had been very close to the perfection God had intended, but thereafter, as they spread to different parts of the globe and diversified into races, they degenerated. Obviously (to the racist eye), the Blacks were the ones who'd degenerated the most, and a frequent reason given for this was climate: the hotter the climate, the greater the degeneration. Even the liberals of their day could subscribe to this entirely unevidenced pseudoscience, as when Samuel Stanhope Smith (1751–1819), president of what would later become Princeton, wistfully hoped that the Blacks in the US, since they were now exposed to the colder climate, would in due course turn White and be able to take their place alongside their more fortunate compatriots. Georges-Louis Leclerc, Comte de Buffon (1707–1788), a towering figure in the history of French science and a staunch advocate of the abolition of slavery, nevertheless regarded Whites as the superior race:

> The most temperate climate lies between the 40th and 50th degree of latitude, and it produces the most handsome and beautiful men. It is from this climate that the ideas of the genuine colour of mankind, and of the various degrees of beauty, ought to be derived.*

But not everyone agreed that climate could be responsible. The English anatomist William Lawrence (1783–1869), in *Physiology, Zoology and the Natural History of Man* (1828), a book which forcefully put forward the then-startling claim that there were so many similarities between humankind and the rest of the animal kingdom that it was folly to regard human beings as other than animals, noticed that Black Americans, despite having existed in a different climate from Africa's for centuries, had not changed colour. He advanced the idea of domestication as a cause for racial variation, although later in life, seeing how this hypothesis seemed unworkable, opted instead for environmental factors, climate included. Similarly the English anthropolo-

---

* As cited in Stephen Jay Gould's *The Mismeasure of Man* (1981).

gist, anatomist and staunch abolitionist James Prichard (1786–1848), in his *Researches into the Physical History of Man* (1813), proposed domestication as the mechanism but later opted for environmental factors. Interestingly Prichard, because of his abolitionist views, went out of his way to stress that apes and humans were of different lineages, for fear that pro-slavery advocates might claim Blacks were the missing link and thereby justify their abhorrent slave-owning ways.

An alternative explanation for the perceived differences between the races was polygeny, the notion that the races had been created separately and were in effect different species; the fact that they could and did interbreed made no difference to this argument – after all, couldn't donkeys and horses interbreed to produce mules? If the coloured races weren't true races of mankind but really just humanoid animals, then there was surely nothing morally reprehensible about the Whites, the "genuine" strain of humanity, enslaving them – and indeed some theologians argued that it was the Whites' moral duty to do so, for had not God issued instructions that the beasts of the field were there to serve Man? The latter part of this equation was excoriated memorably by the French anatomist (and non-polygenist) Antoine Étienne Serres (1786–1868) in his *Principes d'Emryogénie, de Zoogénie et de Teratogénie* (1860) as "a theory put into practice in the United States of America, to the shame of civilization". In fact, it had, not all that long before, been the shame of European civilization as well.*

Polygeny and attempts to justify slavery did not necessarily go hand in hand, though. Equally robust in rejecting slavery was the UK physician Charles White (1728–1813), author of the "manifesto" of polygeny: *Account of the Regular Gradation in Man* (1799). White famously included this eulogy to the fortuned race:

> Where shall we find, unless in the European, that nobly arched head, containing such a quantity of brain, and supported by a hollow conical pillar, entering its centre? Where the perpen-

---

* Serres is notorious for arguing that Africans were more primitive than Europeans because the distance between their navel and their penis remained relatively small throughout life, whereas that of Europeans started small but increased as the individual male grew older: the rise of the navel relative to body height was a mark of civilization.

dicular face, the prominent nose, and round projecting chin? Where that variety of features, and fullness of expression; those long, flowing, graceful ringlets; that majestic beard, those rosy cheeks and coral lips? Where that erect posture of the body and noble gait? In what other quarter of the globe shall we find the blush that overspreads the soft features of the beautiful women of Europe, that emblem of modesty, of delicate feelings, and of sense? What nice expression of the amiable and softer passions in the countenance; and that general elegance of features and complexion? Where, except on the bosom of the European woman, two such plump and snowy white hemispheres, tipt with vermilion?

Polygeny's biggest villain was undoubtedly the UK philosopher David Hume (1711–1776), who used the hypothesis as an excuse to promote the most extreme racism. The coloured races should realize and accept their inferiority, and look upon the Whites as their saviours, gratefully accepting colonialism, slavery and in some instances even genocide as blessings bestowed upon them. In his essay "Of National Characters" (printed in *The Philosophical Works of David Hume* [1854]) he wrote:

> I am apt to suspect the negroes and in general all the other species of men (for there are four or five different kinds) to be naturally inferior to the whites. There never was a civilized nation of any other complexion than white,* nor even any individual eminent either in action or speculation. No ingenious manufactures amongst them, no arts, no sciences. On the other hand, the most rude and barbarous of the whites, such as the ancient GERMANS, the present TARTARS, have still something eminent about them, in their valour, form of government, or some other particular. Such a uniform and constant difference could not happen, in so many countries and ages, if nature had not made an original distinction betwixt these breeds of men. Not to mention our colonies, there are Negroe [sic] slaves dispersed all over EUROPE, of which none ever discovered any symptoms of ingenuity; tho' low people, without education, will start up amongst us, and distinguish themselves in every profession. In JAMAICA indeed they talk of one negroe as a man of parts and learning;

---

* Um. China? Egypt? Japan? India? Colombia? . . .

but 'tis likely he is admired for very slender accomplishments,
like a parrot, who speaks a few words plainly.

If we compare this passage with what Hume himself had to
say in the Introduction to his *Treatise of Human Nature*
(1739–40) we don't know whether to laugh or weep: "And as
the science of man is the only solid foundation for the other
sciences, so the only solid foundation we can give to this
science itself must be laid on experience and observation
. . ." When it came to the crunch, as he sought to justify his
irrational hatred of the "inferior species" and his support
for the slave trade that granted himself and his social class
a more luxurious life, Hume cared nothing for his own
complete ignorance of the many civilizations discovered by
Europeans in Africa, Asia and the Americas – or perhaps,
like so many pseudoscientists, he was unwilling to let mere
facts get in the way of his claims.

Hume was hugely influential through the 18th century
and beyond, and it was likely because of his influence that
the 1798 edition of the *Encyclopedia Britannica* bore this
shameful entry, in which, note, Blacks are given a different
taxonomic name from that of true *Homo sapiens*:

> NEGRO, *Homo pelli nigra*, a name given to a variety of the
> human species, who are entirely black, and are found in the
> Torrid zone, especially in that part of Africa which lies within
> the tropics. In the complexion of negroes we meet with various
> shades; but they likewise differ far from other men in all the
> features of their face. Round cheeks, high cheek-bones, a fore-
> head somewhat elevated, a short, broad, flat nose, thick lips,
> small ears, ugliness, and irregularity of shape, characterize
> their external appearance. The negro women have the loins
> greatly depressed, and very large buttocks, which give the back
> the shape of a saddle. Vices the most notorious seem to be the
> portion of this unhappy race: idleness, treachery, revenge,
> cruelty, impudence, stealing, lying, profanity, debauchery,
> nastiness and intemperance, are said to have extinguished the
> principles of natural law, and to have silenced the reproofs of
> conscience. They are strangers to every sentiment of compas-
> sion, and are an awful example of the corruption of man when
> left to himself.

Eric Morton, who cites this passage and others reproduced
here in his essay "Race and Racism in the Works of David
Hume" (2002), adds: ". . . and the editors continued

publishing entries in this vein until well into the twentieth century."

With such attitudes being, if not universal, at least widespread in the 19th century and even after, novels like H. Rider Haggard's *King Solomon's Mines* (1885) and John Buchan's *Prester John* (1910), in both of which at least some Blacks are regarded as equal or superior to the finest the Caucasian race has to offer, must have landed like a bombshell in the average Victorian or Edwardian living room. And we should not forget Kipling's "You're a better man than I am, Gunga Din". How much greater must have been the impact in the US, where the influence of this and similar racist pseudoscience was far slower to wane than in Europe. There one found anthropologists of the highest distinction spouting what even at the time must have seemed baloney to their European colleagues. No wonder Harriet Beecher Stowe's *Uncle Tom's Cabin* (1852) was so startling.

One of those US anthropologists was in fact Swiss-born: Louis Agassiz (1807–1873). A disciple of Cuvier, Agassiz made his name while still in Europe as a palaeontologist and glaciologist; when he decided to emigrate to the US in 1846, he was heralded as the giant of science he looked set to become. He was appointed Professor of Natural History at the Lawrence Scientific School, Harvard, in 1847, a position he held for the rest of his life. Agassiz seems to have shown no particular signs of militant racism before he came to the US, almost certainly because he had never met any Black people up to then; in his writings he records his immediate gut revulsion on meeting Blacks for the first time, and this primitive reaction proceeded to engender in him a profound detestation of the coloured races, Blacks in particular. The idea of interbreeding between Blacks and Whites seems to have become an outright phobia: "The production of half-breeds is as much a sin against nature as incest in a civilized community is a sin against purity of character." (To his credit, it should be added that at the same time he vociferously opposed slavery.) Of course, as a scientist, he quickly sought some means, however desperate, of justifying his purely irrational emotion.

Polygeny was the obvious answer. It was, as an added bonus, in good accord with his earlier work, because Agassiz was a taxonomist with a vengeance. Where other taxonomists would note the strong similarities between widely distributed creatures and realize they were all merely different races of the same species, Agassiz would blithely name each variant, however minor the distinction, as a separate species. In order to make sense of his prodigious species-generation, he had evolved the hypothesis that creatures were

Louis Agassiz

brought into being at various "centres of creation", near which they tended to stick; he very much downplayed the possibilities of large-scale migration from a species' point of origin. Thus the idea that the various races of mankind were distinct, separately created species seemed a natural one – although there was one problem: Agassiz, like many other scientists of his time (this was before 1859 and the explosion of Darwinism), was a devout Creationist, and it was hard to equate polygeny with the notion that we were all descended from Adam and Eve. However, he rationalized himself out of this dilemma by deciding that *Genesis* spoke only of the region of the world known to that book's writers; beyond their knowledge could have been the creation by God of multiple Adams and Eves elsewhere around the globe – "centres of creation" again.

This hypothesis would not in itself necessarily have fostered racism; however, Agassiz felt driven to add to it, basing his extrapolations on his own profound ignorance of the races of Man:

> It seems to us to be mock-philanthropy and mock-philosophy to assume that all races have the same abilities, enjoy the same powers, and show the same natural dispositions, and that in consequence of this equality they are entitled to the same position in human society.

There was no point in educating Blacks for other than hard labour; in his astonishingly unscientific ranking of the moral and intellectual qualities displayed by the various races, Blacks were at the bottom of the pile in every possible category, and that was the place they should occupy in society, too. Reading some of his writings on the subject of the races – there's a good selection in Stephen Jay Gould's *The Mismeasure of Man* (1981), along with a fuller account than here of Agassiz's thinking – it's difficult to conceive how a supposed man of science was able to persuade himself that what he was propounding was science, or anything like it: most of his "evidence" has the status of an old wives' tale.

An important contemporary of Agassiz in the polygeny field was the US physician Samuel George Morton (1799–1851). He early on adopted the principle of polygeny and its "natural" extension: that the races could be placed in order of rank – the usual Whites-at-the-top-and-Blacks-at-the-bottom ranking, of course. In order to prove his hypothesis, he amassed a huge collection of human skulls – hundreds – whose cranial capacities he measured as an indication of brain size, filling the skull cavities with sifted white mustard seed and then pouring the seed into a measuring cylinder. Surprise, surprise, he found this measure supported his notions, and he published extensive tables to back up the contention. The differences between the average cranial capacities seemed dramatic, and his research was for decades regarded as conclusive. Much later, however, Stephen Jay Gould – again see *The Mismeasure of Man* – went through Morton's research with a fine-toothed comb and discovered the dramatic extent to which Morton had, most likely subconsciously, tailored his results in order to fit with his hypothesis. Re-analysing Morton's own raw data, Gould discovered that they indicated only minor differences between the races – minor enough indeed that there's no particular reason to believe other than that they're a mere statistical effect. Several hundred skulls may seem a large sample to have measured – and of course in a way it is – but as a representation of humanity's billions it is infinitesimally small.

Morton's polygenistic flag was kept flying by his student, Josiah Nott (1804–1873), and the Egyptologist George Gliddon (1809–1858) in their stunningly racist and profoundly pseudoscientific *Types of Mankind, or Ethnological Researches* (1854); much of Gliddon's Egyptological work has been, and was even at the time, regarded as similarly scientifically suspect.

It might be thought that, with the advent of our understanding of evolution, human evolution included, and the publication of Charles Darwin's *On the Origin of Species* (1859), the idea of mutually inferior and superior races would ebb dramatically. However, evolution too can be used as a justification for racism if the evolutionist is so inclined. Even Darwin himself was prone to misinterpret the consequences of his own theory, *inter alia* forgetting (in *The Descent of Man* [1871]) that one of the factors in future human evolution is likely to be human free will:

> At some future period, not very distant as measured by centuries, the civilized races of man will almost certainly exterminate, and replace, the savage races throughout the world. . . . The break between man and his nearest allies will then be wider, for it will intervene between man in a more civilized state, as we may hope, even than the Caucasian, and some ape as low as a baboon, instead of as now between the negro or Australian and the gorilla.

Darwin clearly had a very Eurocentric view of the meaning of the term "savage races". And T.H. Huxley (1825–1895), "Darwin's Bulldog", was little better, writing in 1871:

> No rational man, cognizant of the facts, believes that the average negro is the equal, still less the superior, of the white man. And if this be true, it is simply incredible that, when all his disabilities are removed, and our prognathous relative has a fair field and no favour, as well as no oppressor, he will be able to compete successfully with his bigger-brained and smaller-jawed rival, in a contest which is to be carried out by thoughts and not by bites.

It's interesting to notice that Huxley was yet another on whom weighed heavily the notion that the supposed over-prognathosity of the African skull was an indication that Africans had evolved for eating and biting rather than, as per their European cousins, intelligence. Again, it's hard to comprehend how Huxley, of all people, did not realize that, even if the speculation were true concerning the jaw size being an adaptation toward eating and biting, this was totally irrelevant to the adaptation for intelligence.

Just because Darwin had changed our views on human ancestry didn't mean there weren't hold-outs. Among these was the Edinburgh anatomist Robert Knox (1791–1862), known primarily to us today as the best customer of the serial murderers Burke and Hare and subject of James Bridie's play *The Anatomist* (1931). His association with the murderers, although he was publicly exonerated of any guilt or complicity, led to his widespread ostracism from Edinburgh society, and in 1842 he moved to London. It was there that he began his anatomical studies of the peoples of Southern Africa, developing and becoming near-obsessed by ethnological hypotheses that lacked all evidential support. To judge by the absence of scientific objectivity in this passage he must have suffered some kind of mental collapse after his experiences in Edinburgh:

> Look at the Negro. Is he shaped like any white person? Is the anatomy of his frame, his muscles or organs like ours? Does he walk like us, think like us, act like us? Not in the least! What an innate hatred the Saxon has for him!

What's truly frightening is how easy it is to find people in the 21st century who say very much the same thing, as if it formed some sort of rational argument. Knox's student James Hunt (1833–1869), a profound racist, founded the Anthropological Society of London in 1863 to advance his polygenistic views. An important early member was the ethnologist and translator Richard Burton (1821–1890).

For a long time, as we saw, it was believed the three races of Man – European, African and Asiatic – must reflect the three sons of Noah who had survived the Flood: Japheth was the paternal ancestor of the European race, Ham of the Africans and Shem of the Asiatics. Columbus's opening up of the New World, and the discovery that the inhabitants of it were of a fourth, hitherto-unknown race, threw the standard model into some disarray.

One hypothesis advanced – by, for example, Isaac de la Peyrère (1596–1676) in *Men Before Adam* (1656) – was that the Native Americans, along with all other coloured people, were the descendants of a separate, earlier Creation of Man from the one in the Garden of Eden; these people, having been created at the same time as the beasts, were really likewise to be regarded as beasts. Somehow they had escaped drowning in the Flood, so their descendants were still alive, co-existing with "true" human beings, the descendants of Adam and Eve. De la Peyrère also proposed that *Genesis* was not really a history of humankind as a whole, just a history of the Jews. But these notions were a little too daring for their day, suggesting as they did both that the account in *Genesis* was incomplete and that all good, God-fearing Christians were in fact Jews. He was thrown in prison for his dangerous ideas.

The general conclusion was that the Native Americans must have colonized the Americas from *somewhere else*, and the debate began as to where that somewhere else might have been.

Bizarrely, the true explanation for the origin of the Native Americans – that they had migrated from northeastern Asia to the northwest of the Americas, and spread southward from there – was one of the earlier ideas advanced, purely on the basis of the physiological resemblances between Native Americans and Asiatics, such as the broadness of the face and most especially the epicanthic fold of the eye. The Italian navigator Giovanni da Verrazano (or Verrazzano; c1485–1528) came to this view after his famous 1524 voyage in which he became the first European to see Manhattan Island; he did not land there, but he landed in various other places and noticed the Asianness of the natives' features. The Portuguese seafarer Antonio Galvaño (d1557), author of *Discoveries of the World* (1555) and Governor of Ternate, arrived at the same conclusion, and for the same reason, about 30 years later.

But the best exposition of this hypothesis was undoubtedly that by the Jesuit friar Joseph de Acosta in his *The Natural and Moral History of the Indies* (1590). Acosta noted the same facial similarities as others before him, but he went far further. He observed that many of the wild animals found in the Americas strongly resembled their counterparts in the Old World. He reasoned that, while immigrant humans might have brought with them their domesticated animals, they would not have brought predators like bears and wolves. The only conclusion must be that the animals had been able to reach the New World under their own steam – i.e., on foot. People had presumably done likewise.

Acosta therefore speculated there must be a place where the New World met the Old, that the two landmasses could not be entirely separate. This had to be in some today little-frequented area, and he proposed that in the far northeast of Asia there must be a crossing-place allowing ingress to the northwest of the American continent. Not until much later, in 1648, was the narrow Bering Strait discovered, by the Russian explorer Semyon Dezhnev (c1605–1673), and in fact it wasn't until the 1741 expedition of the Danish navigator Vitus Bering (1681–1741) that the world became aware of how close the two landmasses approached each other. Acosta's deductions were thus over a century and a half before their time.

But these rational voices were drowned in the chorus of wilder speculations. The Lost Tribes of Israel were a frequent choice as ancestors for the Native Americans, especially after the Dutch scholar Manasseh ben Israel (1604–1657) published his popular book *The Hope of Israel* (1650). The idea originated far earlier than this, however, possibly with a suggestion by the Spanish priest Diego Duran (c1537–1588) in 1580. In a very different context, it was regurgitated to dramatic effect in the writings of the founder of Mormonism, Joseph Smith (1805–1844). The claim in the Book of Mormon that Native Americans were the Lost Tribes was to cause significant embarrassment to the Church of Jesus Christ of Latter-Day Saints over 150 years later when testing of the genomes of Native Americans during the 1990s proved beyond all possibility of doubt that they were of Asiatic descent; how could God have got it so wrong when transmitting the information to Smith? At the start of the 21st century the more liberal wing of the Church were calling upon their leaders to apologize to the millions of Native Americans it had converted under false pretences, as it were; far from agreeing, Mormon leaders were describing as heretical the very notion that the DNA studies undermined the validity of the Book of Mormon, and were presenting the public face that the genetics and the scripture were not in fact mutually incompatible. Mormons trying to strike a middle course put forward such hypotheses as that the Lost Tribes first went to Asia, where their genes were "swamped" through interbreeding with the

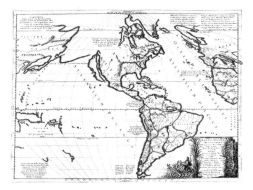

America as the real Atlantis – a map from 1670

locals; it was sometime thereafter that the hybrid population came to the Americas.

Even so, such ameliorative efforts did not stop the Mormon leaders from beginning excommunication proceedings against a Mormon anthropology professor, Thomas W. Murphy of Edmonds Community College, Washington State, who came out into the open in 2002–3 about the difficulty in believing the historical veracity of the Book of Mormon in light not just of the DNA results but also of its claims that darkness of skin indicated sinfulness and that a new human life has its sole origin in the father's sperm. The case attracted widespread media attention – some dubbed Murphy the "Galileo of Mormonism" – and, rather than remain in the limelight of publicity, the Church backed down.

Also popular as a source for the Native Americans was Atlantis. This was the belief of, among many others, the Spanish cleric Francisco Lopez de Gómara (1510–c1566), who in 1552 presented an argument based on the fact that there is a word, *atl*, in the Nahuatl language that means "water": what clearer reference could there be to an Atlantean origin? The Dominican Gregorio Garcia (*fl*1600) was another who favoured the Atlantean hypothesis, according to his book *Origins of the Indians* (1607); on the other hand, in the same book he also proposed they descended from voyaging Greeks, Chinese, Vikings and/or others, not to mention their being the Lost Tribes of Israel – he described the Native Americans as craven Jews, fit only for manual labour – so who can tell how much credence he put in the Atlantean version?

Astonishingly, the Atlantean hypothesis still had its adherents in the latter part of the 20th century: in his book *Psychic Archaeology: Time Machine to the Past* (1977) Jeffrey Goodman, basing his claims on paranormal methods, predicted that a particular site in Arizona would soon reveal indisputable evidence that it was first populated by refugees from the Atlantis catastrophe. Sadly his prediction was not fulfilled.

The Native Americans themselves didn't help much in the sorting out of all this muddled thinking, for their own myths-of-origin declared they were immigrants from nowhere, being created in the very territory they still inhabited. Again, this thesis is still being presented on occasion. Jeffrey Goodman, once more, proposes in *American Genesis* (1981) not only that the Native Americans were the creators, in prehistoric times, of the first human civilization but that it was from the New World that civilization then spread to the Old. Vine Deloria Jr, in his *Red Earth, White Lies: Native Americans and the Myth of Scientific Fact* (1995), roots his arguments for a separate North American human Creation more in the indigenous myths and legends than in the kind of psychic means favoured by Goodman. He also argues that the onetime existence of the Bering Landbridge is merely a matter of supposition, not proven fact. It would appear Deloria's motivations are primarily political.

In 1824 Antoine Fabre d'Olivet (1767–1825) published his *Histoire Philosophique du Genre Humaine*, in which he announced a brand-new theory of human evolution. Rather than consider that there is a single species of Man, whose races may have different colours and cultural histories as well as minor physiological differences but are otherwise of similar nature and antiquity, he theorized that the various races – distinguished by their colours – have succeeded each other. Thus the Native Americans are the relics of the primordial Red human race, largely extinguished when Atlantis was lost; then came the Blacks (although there seem rather a lot of survivors of this race); and most recently the Whites. No mention in d'Olivet's scheme of the Yellows and Browns.

While the hypothesis is ridiculous – and was so even at the pre-Darwinian time – this did not deter the Theosophists from adopting large parts of d'Olivet's theory. According to Madame Helena Blavatsky, founder of the Theosophical Society in 1875, there will be in all seven root races of Man, of which we are the fifth. Each of these has seven subraces, and it is from the subraces of one that the next root race is born: the original root race perishes as its continent sinks. The first and second root races were, respectively, totally aethe-real and partly aethereal; it is unclear, in this light, why they should have drowned as their continent sank. The third root race inhabited Lemuria: their continent was inundated when they committed the sin of discovering sex. The fourth root race occupied Atlantis: its survivors became the Mongolians. They got into the habit of miscegenation, so their offspring were monsters. The descendants of these monsters are still visible among us as the "lesser races" of mankind. So far our own root race, the Aryans, has produced only five subraces out of its allotted seven, but a sixth distinct subrace is apparently now making its appearance in California, so our time is running out.

The evidence in favour of the root-race theory is all around us, of course. As Blavatsky pointed out, the Easter Island statues definitely depict members of one of the other root races – probably the Atlanteans.

Another of Blavatsky's more bizarre hypotheses concerning human evolution, expressed in her book *The Secret Doctrine* (1888), reversed conventional wisdom by saying that in fact apes descended from Man. The Mongolians, the previous root race, had been replaced by the Aryans. Apes owed their origin to inbreeding among the Mongolians. She extended this reasoning to claim that all mammalian life owed its origin to the human variety.

While such ideas seem ludicrous to us now, we should realize they were merely part of a spectrum of 19th- and early-20th-century racist pseudoscience, and by no means at the most extreme fringe of that spectrum. A titan of society like Thomas Alva Edison was one of the early members of the Theosophical Society; even though he remained so but briefly, he clearly didn't find their ideas entirely ridiculous or he'd never have joined in the first place. With so much nonsense being so widely accepted in even the topmost social strata, it's hardly surprising that a movement like eugenics sprang up.

EUGENICS

Although often perceived to be a pseudoscience or even perhaps scientifically oriented, eugenics was and is not so much either of these as a belief system. The idea that the species can be improved either through the prevention of "undesirables" breeding (negative eugenics) or the encouragement of breeding between "ideal" partners (positive eugenics) dates back long before any ideas of evolution came onto the scene – Plato (427–347BC) mentioned notions along these lines – but really came to prominence after the publication of Darwin's *On the Origin of Species* in

1859. A prime early advocate was a cousin of Darwin's, Francis Galton (1822–1911), who also coined the term "eugenics". His statistical researches into human heredity convinced him that intelligence and other qualities – such as courage and honour – were inherited characteristics, and this encouraged him to become a proselytizer on behalf of "racial improvement" through both positive and negative selection. Galton's ideal of the human species was the Anglo-Saxon model; on the European continent, like-minded thinkers opted for the Nordics, supposed descendants of the once-great Aryan race such theorists near-worshipped.

In the US, the idea of eugenics fell on fertile soil. People of every nationality, creed and colour were arriving in the country, while the end of the Civil War caused a further mixing of populations as the poor from the South sought a living in the North; in places the melting-pot effect was successful, but often, as in so many instances where communities feel vulnerable, people sought scapegoats whom they could blame and, best of all, regard as inferior human beings. Racism was rampant in a complex of forms, as was religious discrimination, and eugenics, with its "scientific" veneer, was perfectly tailored to be a socially acceptable way of expressing these hatreds. In particular, since the eugenicists believed Blacks were a separate and inferior strain of humanity with a distinct evolutionary history, obviously miscegenation was to be discouraged as deleterious to the White lineage. (Strangely, no one appeared to wonder why all the best and most intelligent dogs seem to be mongrels.) The ideal scapegoat group was relatively powerless in society, so the immigrant Irish Catholics were a good choice; in turn, the Irish blamed and hated the Blacks, the single easily identifiable group in society who had even less power than the Irish; and so on.

The most prominent eugenicist in the US was the lawyer and Aryan aficionado Madison Grant (1865–1937), author of the books *The Passing of the Great Race* (1916) and *The Conquest of a Continent* (1933), among others. A marked success of Grant and his cronies was the passage by the Federal Government of the Johnson/Reed Act of 1924 (not repealed until 1952) that selectively restricted immigration, with people of "undesirable" stock being discriminated against. As well as the restriction of immigration, Grant advocated racial segregation (primarily as a way of avoiding miscegenation) and the sterilization of members of "inferior" races as well as "feeble-minded" Whites. In 1907 Indiana passed a state sterilization law, and 31 other state legislatures followed suit over the next two decades. The definition of "undesirable" could be very broad indeed: under the 1913 Iowa state law, "criminals, rapists, idiots, feeble-minded, imbeciles, lunatics, drunkards, drug fiends, epileptics, syphilitics, moral and sexual perverts, and diseased and degenerate persons" were all eligible for enforced sterilization. Even so, few of the states in fact carried out many sterilizations, since the popular mood was not wholly in favour and since there were questions as to whether the operation constituted "cruel and unusual punishment".

In 1927, however, the *Buck vs. Bell* case came before the Supreme Court. The teenage mother Carrie Buck – deemed feeble-minded solely because she had conceived out of wedlock – had been placed in the Virginia Colony for Epileptics and the Feebleminded, and the State of Virginia was determined to sterilize both her and her child on the grounds they were a drain on the state economy and their offspring would be a further drain. In the ensuing court case, the state enlisted the "scientific" support of the Eugenics Record Office, a laboratory founded in order to research ways of "improving" the US population by Grant ally Charles Davenport (1866–1944), author of such books as *Heredity in Relation to Eugenics* (1911), and funded by the Carnegie Institution. The ERO's testimony was weighty enough that the case reached the Supreme Court – where, shamefully, it was found in favour of the State of Virginia. The degree to which the judges bought into the eugenicists' pseudoscience can be assessed by the majority statement of Judge Oliver Wendell Holmes (1841–1935), son of the great essayist:

> We have seen more than once that the public welfare may call upon the best citizens for their lives. It would be strange if it could not call upon those who already sap the strength of the state for these lesser sacrifices, often not felt to be such by those concerned, in order to prevent our being swamped with incompetence. It is better for all the world, if instead of waiting to execute degenerate offspring for crime, or to let them starve for their imbecility, society can prevent those who are manifestly unfit from continuing their kind.

The judgement was the green light for enforced sterilizations, which during the 1930s averaged 2200 per year in the US; by 1945 over 45,000 people had been compulsorily sterilized, of whom about half were inmates of state mental institutions. Almost half of all these operations were carried out by the State of California. The consequences of the Supreme Court decision were not just national, however, but

– and tragically so – international. Over the next few years, sterilization laws were passed in Denmark, Norway, Sweden, Finland and Iceland.

And then there was Germany. There was a strong US connection to the ghastly happenings there, too – a connection that US historians not unnaturally tend to gloss over. In *Mein Kampf* (1925) Adolf Hitler (1889–1945) promoted eugenics-based sterilization heavily, and in that same year the US Rockefeller Foundation gave $2.5 million to the Munich Psychiatric Institute as well as further money to Berlin's Kaiser Wilhelm Institute for Anthropology, Human Genetics and Eugenics, all in order to promote eugenics-oriented research. On Hitler's accession to power in 1933, one of the first acts his government passed was a sterilization law; the onus was placed upon physicians to report to a Hereditary Health Court any time they came across someone who was "deficient". The German law was to a large extent based on the existing law in California. In the following year, the American Public Health Association publicly praised the German law as a prime example of good science-based health policy that would benefit society, while the *New England Journal of Medicine* and even the *New York Times* – a strong supporter of the US sterilization laws – were effusive in their approval. By 1940, nearly 400,000 Germans had been sterilized according to the country's law. Rather than being horrified, US eugenicists were concerned their nation was lagging behind Germany's sterling example, and urged greater efforts to catch up.

Although Germany from about 1940 onward took the notion of simply murdering the insane and other "defectives" to an unenvisaged extreme, once again the idea was a product of the eugenics fanatics in the US, in particular Madison Grant's 1916 book *The Passing of the Great Race*, of which Hitler was a great fan. Grant's position was that, if killing the unfit was the only way to stop them breeding, then it was preferable to allowing them to stay alive. Grant was far from alone in this view – and far from the most extreme. Alexis Carrel (1873–1944), winner of the 1912 Nobel Physiology or Medicine Prize and employed by the Rockefeller Institute for Medical Research, wrote in *Man the Unknown* (1935):

> Gigantic sums are now required to maintain prisons and insane asylums and protect the public against gangsters and lunatics. Why do we preserve these useless and harmful beings? The abnormal prevent the development of the normal. This fact must be squarely faced. Why should society not dispose of the criminals and insane in a more economical manner?

His answer was that they should be "humanely and economically disposed of in small euthanasic institutions supplied with proper gases". That was advice the Nazis took to heart; between 1940 and 1941, when Hitler discovered a new, anti-semitic use for his chambers, the Nazis slaughtered some 70,000 of the mentally ill, mainly in Poland, and mainly for the sake of saving money.

After the end of WWII, when the full horrific scale of the atrocities at the German death camps became known to the US public, the ideas of the eugenicists, including enforced sterilization – which at times had enjoyed a 66% approval rating among that same public – took an abrupt nosedive, and fortunately they have remained in a fringe position ever since.

One of the studies that helped fuel the eugenics movement in the US was the book *The Jukes: A Study in Crime, Pauperism, Disease and Heredity* (1877) by Richard L. Dugdale. Dugdale was a volunteer inspector for the New York Prison Association, and in 1874, while visiting the prison of Ulster County, in New York State's Hudson Valley, he noticed that no fewer than six of the prisoners there were blood relatives. Intrigued, he probed further, and discovered that, of 29 male blood relatives, 15 had been convicted of crimes – an extremely high rate for any extended family (well, outside the Mafia). He then traced the family concerned, the Jukeses, back as far as he could in the Hudson Valley, identifying 709 Jukeses under a diversity of surnames, and finally reaching an ancestor called Max, who was born sometime around 1720–40. The branch of the family that had caused all the trouble had begun with a daughter-in-law of Max's, a woman Dugdale called "Margaret, Mother of Criminals". (All the names he used, including "Jukes", were pseudonyms.) Tracing her descendants, he was able to show the family displayed an extraordinarily high incidence of criminality, mental defect and the like, and calculated the financial burden they'd placed on society as an astonishing $1.3 million (equivalent to about $21 million today). He speculated as to whether this catalogue of miscreancy and misfortune was due to heredity – "bad blood" – or environment, coming to no firm conclusion but tending to believe the responsibility was, as it were, the inheritance of a bad environment by each generation from the one preceding it: in other words, the Jukes kids always had a lousy upbringing.

The burgeoning eugenics movement ignored Dugdale's tentative conclusions about environmental influences and declared the Jukeses' failings to be exclusively hereditary, seizing on the Jukes family as an example of the kind of people who might justifiably be weeded out of society's benefit, either by sterilization or by euthanasia. In 1911 Dugdale's original notes were discovered and sent to the Eugenics Record Office. A researcher called Arthur H. Estabrook (1885–1973) was given the job of updating the study. Over the next few years Estabrook claimed to have tracked down a further 2111 Jukeses; there were 1258 alive at the time of his researches, and many of them were – horror of horrors – reproducing to produce yet *more* Jukeses, at vast potential cost to the taxpayer. In his book *The Jukes in 1915* (1915) Estabrook estimated this further cost at over $2 million (over $35 million in today's terms). Estabrook's researches did show, however, that the Jukeses were becoming less problematic – a point the Eugenics Record Office, in its official pronouncement, blithely ignored. At the 1921 Second International Congress of Eugenics, held in New York at the American Museum of Natural History, a full display was devoted to the Jukeses as prime targets for eugenic removal. Right up until the general demise of the US eugenics movement at the end of WWII, the Jukeses were made an example of the kind of problem "sensible" eugenics could cure.

In 2001, however, a poorhouse graveyard was unearthed in Ulster County, and some of the graves there were discovered to be of members of the Jukes clan. Further, some of Estabrook's papers became available to researchers, including his charts of the pseudonyms he and Dugdale had used for the various individuals in the extended family. It

An unidentified
Jukes in
stereotypical pose

emerged that, while indeed there had been plenty of bad
hats in the lineage, there had also been some pillars of society, a fact neither Dugdale nor Estabrook had thought
worth noting. Further, it appeared the real problem beset-
ting the Jukeses was in most instances just straightforward
poverty, which had the effect not only of, in the usual way,
enticing or forcing some family members into criminality
but also of making others vulnerable as scapegoats. The
eugenicists' idea that the family suffered an inheritable
biological flaw was simply untenable, as one might gather
from the title of a recent book on the subject, *The Unfit: A
History of a Bad Idea* (2001) by Elof Axel Carlson. This
hasn't, of course, stopped some modern-day self-appointed
moral arbiters pointing to the family depicted in Dugdale's
and Estabrook's studies as a classic example of the way in
which "immorality" can be inherited. It would appear the
Jukeses' role as scapegoats is not over yet.

THE MORAL COMPASS

Sylvester Graham (1794–1851) was a crusading dietician
and vegetarian in the US; the sugary, overprocessed object
today known as the Graham Cracker is a bastard descendant
of the nutritious wheat-rich biscuit he devised and recommended. A Presbyterian minister, he published books not
only on diet and vegetarianism but also on religion and
morality. In his 1834 work *A Lecture to Young Men on Chastity,
Intended Also for the Serious Consideration of Parents and
Guardians* he let rip a full-scale tirade of anti-sexual pseudoscience:

Those LASCIVIOUS DAY-DREAMS, and amorous reveries, in which young people too generally, – and especially the idle, and the voluptuous, and the sedentary, and the nervous, are exceedingly apt to indulge, are often the sources of general debility, effeminacy, disordered functions, and permanent disease, and even premature death, without the actual exercise of the genital organs! Indeed! This unchastity of thought – this adultery of the mind, is the beginning of immeasurable evil to the human family. . . .

Beyond all question, an immeasurable amount of evil results to the human family, from sexual excess within the precincts of wedlock. Languor, lassitude, muscular relaxation, general debility and heaviness, depression of spirits, loss of appetite, indigestion, faintness and sinking at the pit of the stomach, increased susceptibilities of the skin and lungs to all the atmospheric changes, feebleness of circulation, chilliness, head-ache, melancholy, hypochondria, hysterics, feebleness of all the senses, impaired vision, loss of sight, weakness of the lungs, nervous cough, pulmonary consumption, disorders of the liver and kidneys, urinary difficulties, disorders of the genital organs, weakness of the brain, loss of memory, epilepsy, insanity, apoplexy, – and extreme feebleness and early death of offspring, – are among the too common evils which are caused by sexual excesses between husband and wife. . . .

In other words, sex – marital or extramarital, not just real sex but, even more terrifyingly, merely thinking about it – rotted the mind and body to almost equal, but certainly enormous, degrees. To many readers of Graham's day this must have seemed nothing more than sound commonsense, and a suitably dire warning to wave in the faces of the lascivious, ever impressionable young, or not so young. What's truly alarming is that among some communities these or similar ideas are still taken seriously today, at least as a matter of public morality, even if private standards of behaviour are significantly otherwise. When such attitudes become moulders of national and international policy, and are presented as if the conclusions of science, then we need to be severely worried. Exactly this has happened during the first few years of the 21st century with the Bush Administration's policies on the twin problems of teenage pregnancy and HIV/AIDS.

Numerous scientific studies have shown that the best way of reducing the number of teenage pregnancies is good sex education (which of course includes education in abstinence, in particular the resistance to peer-group pressure, alongside recognition that abstinence cannot last forever), coupled with an emphasis on widespread access to contraception, such access to be as little hindered as possible. Similarly, education, this time in techniques of "safe sex", lies at the heart of any scientifically based social campaign to reduce the incidence of AIDS, coupled with the easy availability of condoms; condoms cannot guarantee protection from HIV infection (and other STDs), but they can very considerably reduce the risks.

Fundamentalist Christians in the US and elsewhere are generally appalled by the way in which the proven approaches to both problems assume the people involved will actually, you know, *have sex*, and have been active in pushing sexual abstinence as a means of reducing both unwanted pregnancies and, in separate context, the spread of AIDS. The situation in connection with AIDS is further complicated by the fact that the official position of the Roman Catholic Church is to reject all forms of contraception, condoms included; obviously, even if the motive in using a condom is to protect from infection, it's still a contraceptive as well.

Clearly complete sexual abstinence would indeed reduce the risks of both pregnancy and the transmission of STDs, AIDS included, by a full 100%. The tricky part is the actual abstaining. Even those supposedly most enthusiastic about pushing abstinence programs have been revealed time and time again to have difficulty with that half of the equation.

That the problems exist is not at issue: according to 2006 figures, each day in the US alone 10,000 teenagers contract an STD, 2400 become pregnant, and 55 become infected with HIV. What is at issue is the best means of tackling the problems, and here science is clear: abstinence-only programs are ineffective by comparison with the education/contraception mix, and may in fact increase the rate of both pregnancies and STD infections in that, when the resolution of the iron-willed youngsters involved finally cracks, intercourse is likely to be on the spur of the moment and far less likely to involve condoms – assuming the participants can obtain condoms at all. So they risk it, "just this once", and then again, and . . . Among adults, where the primary concern is AIDS, the situation is similar, but worsened by the attitude of especially the males in some communities that condom use is non-macho; this latter is a problem that can only be cured through education. Yet today fully one-third of all the money given by the US for AIDS prevention is spent, and mandated by Congress to be spent, on absti-nence-only programs – i.e., is very largely wasted. Further, in order to pander to the prejudices of the electorally important "moral" "majority",* US funding is systematically withheld from anti-AIDS organizations that advocate condom use.

To put this in some context, many of the victims of AIDS are infants, infected by their mothers at birth or while breast feeding. Some 60% of the infants thus infected die before reaching the age of three. This mortality rate can be cut by half through inexpensive treatment with antiretrovi-ral drugs. In many instances, in poor countries, money is not available for such treatment because it has been chan-nelled away to be used instead on ineffective but "morally" prescribed abstinence-only programs. The victims of this sort of "morality" are thus innocent babies.†

Of course, the victims of this sort of ideological prohi-bition could be the mothers instead. In the early 2000s the pharmaceutical corporation Merck developed a vaccine called Gardasil, with seemingly 100% effectiveness against the most prevalent viruses responsible for cervical cancer; it was accordingly approved in 2006 by the FDA. No wonder. Cervical cancer hits about 14,000 US women each year, killing nearly 4000 of them; worldwide, some 270,000 women died of cervical cancer in 2002. Anything to reduce that nightmarish toll would surely be welcome. Well, not according to religious groups like the Family Research Council, which boasts that it "promotes the Judeo-Christian worldview as the basis for a just, free, and stable society", although presumably the justice and freedom don't extend

---

* Hard to know where to put the quotation marks there, because sacri-ficing other people's lives at the altar of one's own prejudices can hardly be described as moral, while the section of US society concerned is not by any means a majority.
† For more on the deliberate corruption of scientific data by the Bush Administration concerning abstinence-only programs, see pages 310ff.

to the mainly poor women who suffer the highest rates of cervical cancer since they can't afford regular checks for the human papilloma virus (HPV). The FRC's president, Tony Perkins, presented the moral justification for the organization's savage resistance to Gardasil: "Our concern is that this vaccine will be marketed to a segment of the population that should be getting a message about abstinence. It sends the wrong message."

Although the vast majority of people regard syphilis as a scourge, this was in the past by no means a universal opinion. Some, like John Bunyan (1628–1688), regarded it as God's just punishment for mortal sin. Even after Western society had drifted away from the belief that God micromanaged every aspect of the human condition, ideological morality – tritely summarized by the notion that, if it was fun, it must be sinful – still played a part. Scientists were not immune to these misguided ideas. As late as 1860 we find the UK physician Samuel Solly, President of the Royal College of Surgeons, claiming that syphilis was not a curse but a blessing: without the disease as a check, "fornicators would ride rampant through the land" – obviously A Bad Thing. The fallacy in this particular instance of moral ideology attempting to dictate science was that fornicators continued to ride rampant through the land *anyway*; the task of scientists is to deal with the world as it actually is, not the world as they'd like it to be.

Similar prudery was to be found in many of the other medical textbooks of the day.* In John Hilton's *Rest and Pain* (1863) male masturbation – Hilton seems to have been unaware of female masturbation – was accused of causing serious illness and thus something to be prevented at all costs. Hilton's proposed cure was draconian: applying iodine to the penis until it blistered, so that handling it became an agony too far for even the most dedicated masturbator.

Remedies for masturbation were in general even worse than the various awful afflictions supposedly associated with

---

* There's more on the subject in my book *Discarded Science* (2006).

the practice. John H. Kellogg (1852–1943), in his *Embracing the Natural History of Organic Life* (1892), passed on the fruits of his practical experience:

> A remedy which is almost always successful in small boys is circumcision, especially when there is any degree of phimosis [tightness of the foreskin]. The operation should be performed by a surgeon without administering an anaesthetic, as the brief pain attending the operation will have a salutary effect upon the mind, especially if it is connected with the idea of punishment, as it may well be in some cases.

For nonphimosal or already circumcised boys, the prescribed operation "consists in the application of one or more silver sutures in such a way as to prevent erection". Luckily for all concerned, this operation had fallen out of fashion by the era of the airport-security metal detector. But if Kellogg's young male patients should beware, pity the plight of the females:

> In females, the author has found the application of pure carbolic acid to the clitoris an excellent means of allaying the abnormal excitement, and preventing the recurrence of the practice in those whose will-power has become so weakened that the patient is unable to exercise self-control. . . . The worst cases among young women are those in which the disease has advanced so far that erotic thoughts are attended by the same voluptuous sensations which accompany the practice. The author has met many cases of this sort in young women, who acknowledged that the sexual orgasm was thus produced, often several times daily. The application of carbolic acid in the manner described is also useful in these cases in allaying the abnormal excitement, which is a frequent provocation of the practice of this form of mental masturbation.

It seems not to have occurred to Kellogg that any sensible woman would simply stop reporting her "problem" to him.

It comes as something of a shock to find 19th-century prudishness in the medical profession as late as 1959. That was when the British Medical Association withdrew and pulped all quarter-million copies of its annual publication *Getting Married* because an article in it by prominent sexologist Dr Eustace Chesser (d1973) suggested that premarital chastity was nowadays only optional – even though Chesser's conclusion was to recommend it!

Sexual pleasure is not the only sin to have attracted the sharper end of the moral compass. With perhaps better reason, so has booze. The US Prohibition experiment of 1920–33 was a dramatic example of the folly of legislating according to what abstract morality says people ought to do, rather than according to what people actually do.

Perhaps unsurprisingly in that, for decades before it was regarded as a medical condition, alcoholism was dismissed as a sin of the weak-willed, there has been ideology-driven controversy over the condition's treatment. The longstanding approach, as adopted by Alcoholics Anonymous and others, is that alcoholism is a physiological or metabolic disease, possibly congenital, whereby a single drink can trigger the sufferer into uncontrollable drinking. The only treatment, in this view, is total abstinence. There is no such thing as a cured alcoholic, merely an alcoholic who declines to drink. A converse view is that alcoholism is not so much a disease *per se* as a behavioural ailment, and that a better system of treatment involves individual counselling alongside training in the art of controlled drinking.

Leaving aside the debate as to what alcoholism is, the disadvantage of the total-abstinence approach is, obviously, that its prediction can all too easily become self-fulfilling: the abstinent alcoholic, if tempted into that single fateful drink, "knows" there's no escape from plunging back into the habit, and therefore plunges. This is not to say the work of the AA has been without its successes – they are legion – but, equally, there has been a high level of failure.* Any approach that would reduce that failure level would obviously be welcomed by all.

Or not. Various scientific studies have indicated that the counselling/controlled drinking approach does indeed have a higher success rate. One of the fullest of these, published in 1982, was by Drs Linda and Mark Sobell of the Addiction Research Foundation, Toronto: it (and a post-controversy follow-up by the Sobells) showed that after a decade the

* We carefully sidestep the faith-based aspects of the AA method.

traditionally treated group had a mortality rate of about 30% while that of the counselled group was significantly lower (although still appallingly high), at 20%. This paper caused outrage in some circles. Almost immediately a team led by Mary L. Pendery published a rival paper that in effect accused the Sobells of fraud – of fudging the figures.

The consequence was not just one but two investigations of the Sobells' claims – by the Addiction Research Foundation and by the US House of Representatives' Committee on Science & Technology. Both found there was absolutely nothing wrong with the Sobells' work – and most certainly not the slightest suspicion of fraud. Pendery's team had simply set out to attack the conclusion, whatever the validity of the science. In his discussion of the affair in *False Prophets* (1988) Alexander Kohn cites also the physician D.L. Davies, writing in 1981:

> Yet so strong are entrenched ideological views on this issue that the argument waxes even more fiercely, recalling the 19[th-]century battles between wets and drys, using indeed the very language and thoughts of early 19[th-]century temperance workers with the same preoccupation with the morals and religious aspects of the "first drink" and the role of divine help.

In other words, the total-abstinence approach must be the uniquely right one because God says so, whatever the scientific evidence. And, if one is empowered by Divine Authority, any means of attack, no matter how venal, is virtuous. This, of course, makes no sense even in its own terms.

The enforcement of celibacy among the priesthood of the Roman Catholic Church by Pope Gregory VII (c1020–1085) eventually gave rise to a still surviving form of "hidden discrimination" within the sciences.

In the first millennium or so of the Church, celibacy hadn't been much of an issue. It was only with the rise of the monastic movement – initially regarded as an extremist fringe – that the idea of celibacy's desirability came to the fore, and by the end of that millennium it became official Church policy, although only patchily obeyed. Gregory, seeing in priestly celibacy a way of ensuring Church property stayed Church property – celibate priests could have no offspring likely to raise troublesome arguments about inheritance – cracked down harshly. While Gregory's motives were not misogynistic, his move encouraged the development of misogynistic attitudes among the male clergy. When the first universities arrived on the scene around 1200 they did so as offshoots of the cathedral school system, and were for the benefit of the male clergy; the rule of celibacy, and its associated misogyny, thus became institutionalized in them. Women were not admitted,* and thus they missed out – and for some centuries continued to miss out – on the entirety of Europe's renaissance in mathematics and philosophy (the equivalent of science). Such subjects became, in the eyes of society, "not womanly". As late as the 20th century, some Western universities barred women.

We're still suffering the cultural hangover from such attitudes. In the US in the late 20th century only 9% of physicists were female. (A mere 3% of full physics professors were women.) Matters were considerably better in mathematics (36%), chemistry (27%) and especially the life sciences (41%), the latter perhaps being regarded as more "womanly". It is not in any way feminist to hope this social bias will soon disappear: the more good physicists we have, the better for humankind.

The main issue of feminism – that there should be equal opportunities for all, and that something should be done to counter the disadvantages women face – is recognized and accepted by all right-thinking human beings. But not all feminist arguments are entirely convincing. For example, feminists have, on occasion, insisted upon an evolutionary theory which sets them at a higher level than men. In its crudest form, this theory insists that man and woman are of different species. Slightly more subtle is the theory that, while men may well be descended from simian ancestors, women have always been women. This seems equally implausible, although one can understand feminist objections to the account of the origin of woman given in *Genesis*.

---

* Although apparently the rise of the universities closely parallels that of the brothel, it being discovered that students provided a constant and eager clientele.

On the subject of evolution, we should not forget Oscar Kiss Maerth's account in *The Beginning Was the End* (trans. 1973) of the origin of one female fashion. Maerth envisaged bands of primitive, cannibalistic males dashing around gang-raping ape-like subhuman females.

> The rape of the female apes was no simple matter and the mating could not be carried out as is usual among apes, that is, with the male animal entering the female from behind . . . Therefore the females normally had to be pulled to the ground by several cannibal ape-men and laid on their backs so that they could be raped. The legs of the females being raped were raised when they lay in this position and the tips of the feet were extended forward at the moment of orgasm so that the ball of the foot was pushed up. That was a sexually exciting sight for the males taking part in the collective rape and has remained in their subconscious right up to the present. That is why female legs with the tips of the foot stretched downwards and the balls of the feet raised are still sexually stimulating today. This is the origin of high heels.

Elsewhere Maerth remarks that no woman is good at philosophy. He adds grudgingly that, if a woman *is* good at philosophy, there must be something wrong with her sex hormones.

Feminists looking for an evolutionary theory to detest might like to try that of Dr Max Gerson (1881–1959). In 1920 he explained why the menstrual cycle is linked to the moon. Primitive men hunted for their females on nights when the moon was bright. As the women fled they were nevertheless excited by the prospect of what might happen if they were caught, and so experienced a rush of blood to the uterus. Over evolutionary time, this response was exaggerated to become a monthly bloodletting. Gerson didn't explain why something similar isn't exhibited by men.

The hatred for men displayed by some feminists is extreme. Hence *The SCUM Manifesto* (1967; SCUM = Society for Cutting Up Men) by Valerie Solanas (1936–1988):

> The male is a biological accident: the y (male) gene is an incomplete x (female) gene, that is, has an incomplete set of

chromosomes. In other words, the male is an incomplete
female, a walking abortion, aborted at the gene stage. To be
male is to be deficient, emotionally limited; maleness is a deficiency disease and males are emotional cripples.

And Solanas didn't stop there. The male is "trapped in a
twilight zone halfway between humans and apes, and is far
worse off than the apes because, unlike the apes, he is capable of a large array of negative feelings – hate, jealousy,
contempt, disgust, guilt, shame, doubt" . . . which, of
course, women never display. "Although completely physical, the male is unfit even for stud service . . . To call a man
an animal is to flatter him; he's a machine, a walking dildo."
The best remedy for a hapless male might be to go along to
a Turd Session:

> . . . to aid men in this endeavour SCUM will conduct Turd
> Sessions, at which every male present will give a speech begin-
> ning with the sentence: "I am a turd, a lowly abject turd," then
> proceed to list all the ways in which he is.

CORRUPTION OF SCIENCE BY THE
IDEOLOGY OF SCIENCE

Science is itself capable of generating its own scientific
ideologies, of course. As example, there's the idea put about
by Thomas Szasz (b1920) in books like *The Myth of Mental
Illness* (1961) and its successors that there is really no such
thing as mental illness: rather, it is an invention of psychiatrists eager to earn a quick buck by inventing a profession
where none is needed. In Szasz's view, the people we regard
as mentally ill are merely unusual human beings: there is no
ailment there to treat. Insanity is nothing more than a social
construct, decided by the majority. It's rather startling to
find that these very postmodernist-sounding notions –
which seem based on an ideological rather than a scientific
agenda – are being advanced by someone who was for
decades Professor of Psychiatry at Syracuse University. His
position might have seemed more reasonable in previous
centuries, when indeed glorious eccentrics, unless rich
enough, might be classified as mentally ill (as still, of course,

sometimes happens) – or, even earlier, burnt at the stake as witches – but it appears somewhat fantasticated today, when glorious, and even inglorious, eccentrics are more likely to be booked on Oprah or get jobs as Fox News pundits.*

In a quite different area, there are various ideologies involved in the search for extraterrestrial life. Leaving aside the ideologies of ufologists, we're far too prone to regard our own *modus vivendi* as the one likely evolved by organisms elsewhere. In recent decades discoveries on earth have had a rather sobering effect on this preconception: lifeforms have adopted all sorts of ways of coping with environments that would have earlier been thought impossible – and which are certainly far removed from what we'd regard as the terrestrial norm. Recognizing this earth-chauvinism, the National Research Agency now has a panel devoted to "weird life" that advises NASA on what else its Mars missions might look out for rather than earth-type, water-based cells. As was pointed out in 2007 by geologist Dirk Schulze-Makuch of Washington State University, the *Viking* missions of the 1970s, in testing for life, may actually have drowned Martian microbes through the assumption that living cells must be filled with salty water, and thus would respond positively to the addition of a richer water supply. In the environment of Mars, Schulze-Makuch reasoned, where temperatures plummet far below freezing, a much more likely – and perfectly workable – constituent of living cells would be a mixture of water and hydrogen peroxide. Such a mixture does not freeze until reaching about –55°C; moreover, on doing so, unlike water it does not expand, and thus would not necessarily rupture cell walls. NASA's *Phoenix* mission to Mars, to be launched in summer 2007, is planned to be a little more open-minded about the type of lifeforms it looks for than were its *Viking* predecessors.

Today the great talk is of globalization, the process, rapidly increasing in pace, whereby science and technology is permitting – indeed, encouraging – such a large-scale

* That said, it's true that in our modern world vast numbers of people are, thanks to aggressive marketing techniques, being prescribed mind-influencing drugs for which they have no need.

permeation of culture between all the peoples of the world that a single, truly global culture is no longer just a visionary pipedream. At the moment the benefits of globalization are, *pace* prominent enthusiasts like Thomas Friedman (b1953), somewhat intangible: to a great extent the effect has been one of leaching employment from the wealthier countries to the poorer ones, which would be a good thing were it not that the economic priorities of the corporations of the wealthier countries simultaneously seek ever cheaper sources of supply, thereby encouraging the spread of what is effectively slave labour. The shades of those German corporations which discovered before and during WWII (and even, shamefully, for a short while after it) the economic bonus of using slave workers are not so far behind us.

There's a further problem with the emergence of a global culture. Civilizations collapse. History is filled with examples of civilizations that were once the light of the world but then, for whatever reasons, were wiped from the face of the planet. Remember Shelley's *Ozymandias*. (Had Ozymandias thought less of self-aggrandisement and more that cultures require constant effort to maintain themselves, who knows?) Always, however, as one civilization has collapsed there has been another, or at least the potential for another, waiting in the wings: sooner or later after the hiccup, civilization has continued its generally upward path. In the case of a unitary global civilization, though, there can be no such backup: all our eggs will be in a single, very large basket. What will happen if – or, more realistically, *when* – the global civilization collapses?

Science has no answer, for the very good reason that science has yet, aside from a few lonely and distant voices, to address the problem. In large part this is because of the structure of modern science, which structure, while immensely valuable, has an unrecognized corruptive influence on the very science it is supposed to protect and enhance. Science progresses in general through the publication of scientific papers, which are reviewed by other scientists, who accept, reject or amend their conclusions. Modern science is thus primarily an accretional process: complete paradigm shifts, or "scientific revolutions", are an extreme rarity, and anyway are generally found to have had plenty of precursors. Even such a radical paradigm shift as the Darwin–Wallace theory of evolution by natural selection did not come out of nowhere: its precursors can be traced back to ancient times. This accretional, collegial process is the strength of modern science. In general, loony notions do not last long. While science's record in accepting genuine new truths is a bit more patchy, sooner or later those make it into the canon; for example, the theory of continental drift proposed by Alfred Wegener (1880–1930) was rejected and even ridiculed for too many decades until other know-ledge emerging through the collegial system led to its eventual acceptance.

The seriously debilitating weakness of the process is that the measure of scientific success has come to be the publication of those papers. Academic and commercial employers alike, not to mention those who award government grants, want to see a plentiful bibliography attached to the name of any scientist. As we saw earlier, this has led to many cases of fraud, but it is also systemically damaging. The onus is on scientists to publish early and often. The short-term experiment becomes more valuable than the more major one that could take years or even decades to produce a result. Yet the longer-term experiments are very often the ones that are more important for the advancement of human knowledge. When Gregor Mendel, for example, conducted the experiments that laid the groundwork for the whole science of heredity,* he had to observe numerous generations over a period of years. What modern scientist, worried about tenure and with the dean breathing down her or his neck, could consider an experiment that would take that long before a paper could be published?

Of course, some scientists still do. But they do so *in spite of* a system that rewards the frequent publication of trivial knowledge. Perhaps they're placed such that they can ignore the system's imperatives – perhaps they *are* the dean! – or perhaps the experiment is one that, albeit long-term and important, can be done in, effectively, spare time while other things are going on. Fortunately a few commercial entities have noticed the problem, and are beginning to take appropriate steps to solve it: they recognize that, while the bean-counters are clamouring that the important thing

* Though see pages 30ff.

is to sustain profits for the next quarter, it's actually pretty useful to ensure, too, that you'll still have customers in 20 years' time.

This is of course a role that democratic governments used to perform, and many still do. (Totalitarian regimes sometimes think in the longer term too, although often for motives of self-glorification – the Ozymandias syndrome.) Increasingly, however, governments are being lured into the trap of short-term thinking: if the voters are going to the polls in a year's time, you want some Big Result to show them *now*, not just the news that you've started a project that may benefit their grandchildren. An example is the Bush Administration's hogtying of stem-cell research, seeking to capitalize in the short term on the antipathy toward such research of voters on the religious right; blatantly used has been the argument that, while there have been plenty of claims about the potential of stem-cell research, the benefits have not been proven and cannot be so until at least a decade or two down the line. The fact that *this is what good science is all about* is ignored in the rush to appeal, however spuriously, to a powerful voting block.

We're probably right to have some confidence that the scientific establishment, once the problem of short-termism has been properly confronted, will find a way to solve it. At the moment, though, it remains an example of science in effect corrupting itself.

One bizarre form of ideological corruption of science has been nationalism – or, rather, the stereotyping of a country's science along nationalistic grounds. The arch-exponents of this sort of thing were of course the Nazis, whose doomed attempts to create a purely "German science" we'll consider in Chapter 6(*i*). But that was merely one end-point of a story that had been going on in Europe throughout the 18th and 19th centuries, the principal antagonists being Germany and France. Some of the chauvinism was positive (our country's science is the best because . . .) and some of it negative (your country's science is lousy because . . .). As an example of such stupidity, in *The Undergrowth of Science* (2000) Walter Gratzer cites (but alas does not name) the French Minister of Education, speaking in 1852:

Does not our tongue appear especially suited to the culture of the sciences? Its clarity, its sincerity, its lively and at the same time logical turn, which shifts ever so rapidly between the realm of thought and that of feeling – is it not destined to be not merely [scientists'] most natural instrument but also their most valuable guide?

Matters heated up during WWI.* Clearly scientists took sides during that conflict – and some of them served and died – but this was a somewhat different issue than the importation of nationalist enmities into science, which reached such a level that *Nature* was moved to remind its readers that science is not a matter of politics and transitory human preferences.

It would be pleasing to think that such childish follies are behind us, but there are still plenty of cultures, *even in the West*, that regard science as a whole as intrinsically evil simply because it is of Western origin, preferring instead to look to other traditions . . . specifically those that lead one off into woo-woo land. It can certainly be argued that the geography of the area of what we know might have been different had science followed a different cultural course – biology might have advanced at the expense of physics, perhaps – and that this might have been a good or a bad thing, but that's a different *what if?* game. Reality is unaffected by the order in which we discover its secrets. To claim that the nature of reality would somehow have been *other* if we'd gone about the task of exploring it differently is patent nonsense.

One of those who with particular venom denounced Western science as corrupt was Mao Tse-tung. At the time that Marx and Lenin were writing, it was reasonable to believe matter was infinitely divisible, as they said, and that the universe was infinite in both space and time. Since Marx and Lenin must be right in everything they'd written, Mao declared newfangled notions like particle physics, the expanding universe and the Big Bang to be merely the corruptive fictions of bourgeois Western scientists. Despite

* In an amusing prefiguring of recent US attempts to popularize the neologism "freedom fries" for "french fries", there was a move during WWI, again in the US, to rename "German measles" as "liberty measles"!

Mao's strictures, the Chinese managed to develop the atomic bomb and nuclear power, so perhaps the Chairman had two sets of belief, one for public and the other for private consumption.

MEDIA MUFFINS

A further corruption of science that's sometimes ideological, sometimes not, occurs through the fact that almost all of the information most adults receive about developments in science and technology has been filtered through the media – newspapers, magazines, radio and television. While some science journalists are marvellously gifted at their jobs, explaining complexities and significances to the lay audience with a dazzling skill, far too many are not.* Even some of the most prestigious media outlets have science correspondents, sometimes quite well known ones, whose competence as interpreters – and indeed, on occasion, even their grasp on basic science – is distinctly questionable. And all too often one gets the impression that the job of discussing science issues has been fobbed off onto whichever journalist was slowest to leave the room.

Further, some media pundits feel it is their prerogative to make pronouncements on science, and often said pundits display a definite streak of anti-scientism. Richard Dawkins, in *Unweaving the Rainbow* (1998), suggests, probably correctly, that much of this anti-scientism is born from the primitive habit of disliking things we don't understand, and cites several examples of such folly. Here, for instance, is Bernard Levin in *The Times* in 1994 mocking the notion of quarks:

> Can you eat quarks? Can you spread them on your bed when the cold weather comes?

This particular fit of lunacy drew a prompt response, in the form of a Letter to the Editor, from metallurgist Sir Alan Cottrell:

> Mr Bernard Levin asks "Can you eat quarks?" I estimate that he eats 500,000,000,000,000,000,000,000,001 quarks a day . . .

* I'm perfectly aware that I'm in a glasshouse, throwing stones.

On the other side of the Atlantic, anti-scientism is rife among the political pundits, sometimes taking the form of a pretended reverence for scientific studies that prove, to euphemize, a little hard to track down. Here's a December 2006 exchange from *The Radio Factor with Bill O'Reilly* in which O'Reilly (b1949) discusses a significant statistical correlation with his co-host, Edith Hill (b1963):

> *O'Reilly.* 62% of Americans will have a Christmas tree, but most of the trees will be artificial.
> *Hill:* That surprises me. Only 62% have Christmas—
> *O'Reilly:* Yeah. And here . . . and here's a very . . . here's something [they] didn't poll but I know: that most women who like artificial trees—
> *Hill:* Yeah?
> *O'Reilly:* —have artificial breasts.
> *Hill:* What?
> *O'Reilly:* Did you know that? Yeah, there's a correlation. Yeah. There was a study done—
> *Hill:* You know . . .
> *O'Reilly:* It was, it was done at UCLA in LA. All right—
> *Hill:* I don't *believe* you . . .
> *O'Reilly:* We gotta take a break. We gotta take a break, and we'll be back with Reverend Barry Lynn, to talk about why there's so much angst about Christmas . . .

Also from December 2006 there's the *Washington Post* pundit Charles Krauthammer using spurious science to enter the debate concerning the murder of disaffected ex-KGB agent Alexandr Litvinenko (1962–2006) with a dose of polonium-210. Was the murder committed under the auspices of Vladimir Putin (b1952), or could responsibility lie elsewhere? Krauthammer has no doubts:

> Well, you can believe in indeterminacy. Or you can believe the testimony delivered on the only reliable lie detector ever invented – the deathbed – by the victim himself. Litvinenko directly accused Putin of killing him. Litvinenko knew more about his circumstances than anyone else. And on their deathbed, people don't lie.

Not only is it obviously the case that people *do* lie on their deathbeds – "I have always been faithful to you, my darling"

– and that it's a great venue for the settling, honest or otherwise, of old scores, but there's the equal possibility that perfectly truthful people might simply be wrong.

Far more pernicious is a final recent example (chosen from among an embarrassing richness of others), where an ideological agenda either blinds the pundit to established science or dictates straightforward dishonesty. In January 2007, commenting on the freak weather that was devastating the US Midwest, Fox News pundit Neil Cavuto (b1958) commented:

> 24°[F] in Fresno, 29°∞in Phoenix, down to 9°∞in Amarillo, Texas, and on and on. It is some of the coldest air in this part of the country in 20 years. Proof that all of this hype over global warming could be just that – hype?

It is inconceivable that any journalist could be so ignorant of the science concerning climate change that s/he would be unaware that freak extremes of weather – either cold or hot, wet or dry – are possible symptoms of global warming.* Whether he was deliberately corrupting the science to mislead viewers in the hope of promoting his ideology or whether his own comprehension of the relevant science had been corrupted by that same ideology, the result was the same: the scientific understanding *of his viewers* was being grossly corrupted. Had he been as inaccurate about, say, the latest football scores he might well have been fired; as it was, his baloney seemed to go unregistered by his employers – a sad comment on their own attitude toward the relative importance of scientific truth.

The inevitability of the popular media's contribution to the corruption of science may seem obvious: if a piece of science were capable of being explained to the scientifically uneducated within a one- or two-minute news segment, it could hardly have taken countless scientists months, years or even decades to make that particular breakthrough. The

---

* It is equally inconceivable that Cavuto could have been unaware that, at the same time, states like New Jersey were basking in warm, near-summery sunshine rather than suffering their customary early-January snow and ice.

highest the best-intentioned, most responsible of broadcasters can hope for is to present a grossly simplified version,
and a gross simplification of *anything* is by definition a false
account.* Newspapers can make a better attempt, but they
too are trapped in the prison of necessary simplification; for
example, they cannot expect their readers to follow (or their
typesetters to typeset!) page after page of mathematical
equations.

Our hypothetical "best-intentioned, most responsible
of broadcasters" is of course in a minority, at least on TV. Far
more generally, TV is keen to present the most sensational
in science, which very often means stepping right outside
science into the pseudosciences or the downright fraudulent, while still retaining the "science" label.

A further complication, as we've noted, is the popular
media's corruption of the concept of "balance". Balance in
journalism is obviously a good thing: if there is a genuine
dispute in science – as for example there once was between
the Big Bang and Steady State cosmologies – a discussion
between proponents of the conflicting views is likely to be
enlightening; we can at least come away with the correct
impression that there's a debate going on. Where the
corruption occurs is in the many instances where there is *not*
a debate going on but the producers, always in search of
"sexy television", give the viewer the impression there is.
Typically, we're presented with, on the one hand, Talking
Head A and, on the other, Talking Head B. What is withheld is the crucial datum that Talking Head A is a distinguished scientist representing the conclusions of every
researcher in the relevant field while Talking Head B hears
voices. The two opinions are presented as having equal
weight, even though this is a falsehood; if challenged, the
broadcasters are likely to claim piously that they are "leaving it up to the viewer to judge" – which, of course, they're
not, because they've failed to supply the principal basis
upon which the viewer could attempt a judgement.

---

* A few years ago I was offered a commission to write a children's book
on the science of nuclear power that avoided all mention of subatomic
particles, or even that the atom could be split, because subatomic particles were "too difficult". The book was eventually written by someone
else. How much must it have confused its youthful readers?

In this looking-glass world we have, for example, Michael Crichton, whose qualifications are in medicine, presented as an expert on climatology. (In a yet more hilarious example, unqualified demagogue Ann Coulter has been interviewed for her opinions on global warming.) This particular corruption easily translates to political bureaucracies: a feature of the Bush Administration is the appointment as scientific advisors of people on the basis of political loyalty rather than relevant scientific qualification (see pages 295ff and 310ff).

Not just the mainstream media contribute to the corruption of both science itself and the image of science in the popular mind. As noted, a certain amount of public misunderstanding of scientific matters derives from the efforts of popularizers to explain to laypeople material that simply is not explicable in lay terms. But this is a relatively minor effect alongside the efforts of those who, for political or ideological reasons, deliberately essay to distort science in the public mind, sometimes mounting an outright attack in an attempt to destroy the institution that is science: there are plenty of examples in these pages.

But there are also those who set out to corrupt the public understanding of *what science is*, portraying it as "the enemy". Their motivation may be part of an ideological agenda, or it may simply be to gain power or financial profit. In *Sleeping with Extra-Terrestrials* (1999) Wendy Kaminer reserves especial venom for the self-styled gurus and authors of pop self-help/spirituality books – often massively bestselling – and their deliberate falsification and denigration of science. It is a weary rhetorical trick to misrepresent the arguments of one's debating opponent and then attack what was never claimed in the first place, and one would have thought we the public would have wised up to it long ago; yet in "spirituality" as in politics the trick has been used with success for centuries and seems today to be even more effective than ever, as media institutions gullibly or insouciantly promote the dissemination of false ideas. To choose just one of Kaminer's examples: In their Introduction to *The Celestine Prophecy: An Experiential Guide* (1995), James Redfield and Carol Adrienne state that "those who take a strictly intellectual approach to this subject will be the last to 'get it'", and advise readers to "break through the habits of skepticism and denial". As Kaminer summarizes these and many other pop-spiritual authors: "Skepticism they view with contempt, as the refuge of the unenlightened." If you want to read *The Celestine Prophecy*, its own authors advise, you should leave your brain at the door.

Science, in this view, is the enemy of understanding – much as it used to be the claim of bogus spirit mediums that the souls of the departed would fail to materialize should there be anyone in the circle sufficiently ungullible or spoil-sportish to notice the strings and pulleys. What the gurus are essentially saying is: "Only through believing bollocks can you find enlightenment." Of course, the pop gurus don't put it quite like that: those who notice the strings and pulleys are "insufficiently spiritually evolved" or suffer from "closed-mindedness". Those who swallow this stuff whole are, by contrast, the enlightened and open-minded.

One of the leading authors in this field is Dr Deepak Chopra (b1946), whose publications include such notorious works as *Quantum Healing: Exploring the Frontiers of Mind/Body Medicine* (1989) and *Ageless Body, Timeless Mind: The Quantum Alternative to Growing Old* (1993), in which he seems to promote such ideas as that people can cure themselves of cancer by adjusting their own internal quantum mechanics. (This may be a misrepresentation: it's hard to know exactly *what* Chopra means except that there's a lot of quantum involved.) Few of the public and more particularly few of Chopra's eager customers pause to reflect that Chopra himself today looks, well, *older* than he did when these books were published. Of course, such paradoxes are to be expected from a man whose arguments against Darwinian evolution and in favour of Intelligent Design include that there was no such thing as a self-replicating molecule for billions of years after the Big Bang until DNA appeared on earth, without any apparent awareness that there was no such thing as *the earth itself* until billions of years after the Big Bang; if Chopra has a means of detecting the lack of self-replicating molecules in the rest of the universe, he should publish details of this technology in *Nature*.

As has often been remarked, the trouble with having an open mind is that people come along and put their trash in it.

THE CORPORATE CORRUPTION OF SCIENCE

The topic of the negation, suppression, misrepresentation and corruption of science for profit, or to protect profits, by the large commercial corporations is a huge one; there have been a number of books on individual cases alone. No less than the corporations who pay them, the scientists who lend their names to such ventures are corrupting science, either deliberately or unconsciously, although the names of the individuals involved in such dishonesty tend to become less celebrated – if that's the word – than those of the Hwang Woo Suks and the Jan Hendrick Schöns.

Some of the connections are not as obvious as one might think. For example, it's common knowledge that the oil company Exxon extensively funded global warming-denial through the 1990s and on: examples of relevant organizations that receive funding from Exxon include Accuracy in Academia, Accuracy in Media, the Advancement of Sound Science Center, the Advancement of Sound Science Coalition, the Air Quality Standards Coalition, the Alexis de Tocqueville Institution, the Alliance for Climate Strategies, the American Coal Foundation, the American Council on Science and Health, the American Enterprise Institute for Public Policy Research, the American Enterprise Institute–Brookings Joint Center for Regulatory Studies, the American Friends of the Institute for Economic Affairs, the American Petroleum Institute, the Annapolis Center for Science-Based Public Policy, the Arizona State University Office of Climatology, the Aspen Institute, the Association of Concerned Taxpayers, the Atlas Economic Research Foundation, the Capital Research Center, the Cato Institute, the Center for Environmental Education Research, the Center for the Study of Carbon Dioxide and Global Change, the Chemical Education Foundation, Citizens for the Environment and CFE Action Fund, the Clean Water Industry Coalition, the Committee for a Constructive Tomorrow, Consumer Alert, the Cooler Heads Coalition, the Council for Solid Waste Solutions, the

Earthwatch Institute, the Environmental Conservation Organization, the Foundation for Research on Economics and the Environment, the Fraser Institute, the George C. Marshall Institute, the Global Climate Coalition, the Greening Earth Society, Greenwatch, the Harvard Center for Risk Analysis, the Heartland Institute, the Heritage Foundation, the Hudson Institute, the Independent Commission on Environmental Education, the Institute for Biospheric Research, the Institute for Energy Research, the Institute for Regulatory Science, the Institute for the Study of Earth and Man, the James Madison Institute, the Lexington Institute, the Locke Institute, the Mackinac Center, the National Council for Environmental Balance, the National Environmental Policy Institute, the National Wetlands Coalition, the National Wilderness Institute, the Property and Environment Research Center, Public Interest Watch, the Reason Foundation, the Science and Environmental Policy Project, and the Tech Central Science Foundation; then there are the journals *Climate Research Journal* and *World Climate Report*, and the website junkscience.com.* Far less well known, however, is that much of the warming-denial industry is funded not by the oil industry but, as George Monbiot revealed in his book *Heat* (2006), by the tobacco companies.

The tobacco corporations were early entrants to the game of corrupting science, concerned as they were to find

---

* Of course, simply being funded by a commercial operation does not necessarily mean one's publicly expressed scientific conclusions are bogus. However, the partial list given here of Exxon fundees (for a more complete one see www.exxonsecrets.org) represents a measurable percentage of the word's total of scientific global warming-deniers, and it is hard to find the papers the individual scientists have published to back up their public pronouncements. In 2004 *Science* did a survey of 928 randomly selected papers containing the words "global climate change" to find out how many concluded global warming was not largely caused by our burning of fossil fuels. The answer was exactly zero. As one scientist remarked, "There is better scientific consensus on this issue than on any other, with the possible exception of Newton's Laws of Dynamics." Thus not only is the "debate" on climate change monstrously lopsided in terms of numbers, but a high proportion of those in the minority are receiving funding from corporations that misguidedly perceive it to be in their own interests to deny human-generated warming.

"experts" who would claim that smoking – and later passive smoking – was not a contributor to lung cancer. In early 1993 the tobacco company Philip Morris was trying to work out the best way of responding to the 1992 publication of the EPA's massive and damning report *Respiratory Health Effects of Passive Smoking*. They approached a PR company called APCO, which pointed out that the public tended to regard statements on health from tobacco companies with, um, cynicism. APCO proposed, therefore, to set up a *faux*-grassroots organization, a "national coalition intended to educate the media, public officials and the public about the dangers of 'junk science'" and to assail in general the credibility of the science put forward by the US Government's scientific agencies, the EPA included; thus the Philip Morris-generated false science claiming there was no connection between passive smoking and respiratory disease in adults and children would be camouflaged by a welter of false (and perhaps, who knows, genuine) science perpetrated in other areas, all supposedly derived from the work of scientists unconnected with tobacco. This organization became in due course the deceptively titled Advancement of Sound Science Coalition (TASSC). Among the other scientific areas where TASSC proceeded to "cast doubt" upon established, thoroughly peer-reviewed science by pretending there was a difference of opinion in the scientific world was global warming.

Perhaps the single most important element today creating an illusion that there's scientific debate over the reality of climate change is the website www.junkscience.com; in the Orwellian doublespeak characteristic of modern ideological corruptors of science, the website describes genuine scientific research as "junk science" while promoting its own unsupported claims as "sound science". The person who runs the site is Steve Milloy, who started it up in 1992 while working for APCO; in 1997 he became Executive Director of TASSC, which by 1998 was funding www.junkscience.com. In addition, he writes a weekly column, *Junk Science*, for the website run by Rupert Murdoch's flagship cable TV channel Fox News, that essentially regurgitates the contents of the website. Milloy is also apparently responsible for the Free Enterprise Education Institute and the Free Enterprise Action Institute, both of which are funded by Exxon and the latter of which is headed by one Thomas Borelli, previously the executive at Philip Morris who had oversight of the funding of TASSC.

It's a tangled web. Thanks to Monbiot's detective work, we know the spurious science concerning climate change is largely the responsibility of one, scientifically unqualified individual whose funding comes significantly from corporations whose interests would seem quite unrelated.

But there's more. One of TASSC's Advisory Board of eight is Frederick Seitz (b1911), Chairman of the (Exxon-funded) Science and Environmental Policy Project but a scientist distinguished enough that he was President of the National Academy of Sciences 1962–9 and President of Rockefeller University 1968–78. His qualifications are not in climatology or any other of the environmental sciences but in solid-state physics. In 1979 he became a permanent consultant to another tobacco company, R.J. Reynolds, in charge of commissioning research from US universities that might help rebut the damaging health conclusions about cigarette-smoking. Sometime around 1989 he left Reynolds, whose CEO commented: "Dr Seitz is quite elderly and not sufficiently rational to offer advice." Earlier, in 1984, Seitz had co-founded the Exxon-funded George C. Marshall Institute, and it was this organization that in 1994 published his report *Global Warming and Ozone Hole Controversies: A Challenge to Scientific Judgment*. At some stage he became associated with the Oregon Institute of Science and Medicine, a maverick operation founded in 1980 by Arthur B. Robinson (b1942), who started his career as a chemist – under Linus Pauling (1901–1994), no less! – but split off from the mainstream largely, it seems, because his Christian fundamentalism clashed with the findings of real science.

In 1998 the Oregon Institute of Science and Medicine and the George C. Marshall Institute co-published *Research Review of Global Warming Evidence*, written by Robinson and prefaced by Seitz; promoting the madcap notion that increased atmospheric carbon dioxide would bring an era of lush fecundity to the earth – a new Eden – this was produced in a format that exactly mimicked that of the journal *Proceedings of the National Academy of Sciences*, a gambit that mightily confused the media as well as an unknown percentage of the 17,000 graduates (many nonscientists, few qualified to discuss climatology) who signed the accompanying Oregon Petition, which Seitz wrote:

> We urge the United States government to reject the global warming agreement that was written in Kyoto, Japan in December, 1997, and any other similar proposals. The proposed limits on greenhouse gases would harm the environment, hinder the advance of science and technology, and damage the health and welfare of mankind.
>
> There is no convincing scientific evidence that human release of carbon dioxide, methane, or other greenhouse gases is causing or will, in the foreseeable future, cause catastrophic heating of the Earth's atmosphere and disruption of the Earth's climate. Moreover, there is substantial scientific evidence that increases in atmospheric carbon dioxide produce many beneficial effects upon the natural plant and animal environments of the Earth.

Rightly, the Clinton Administration ignored the petition and the *Research Review* as pseudoscientific twaddle; the incoming Bush Administration, however, was able to use it as part of its excuse for withdrawing from the Kyoto Protocol in 2000. State Department papers released in June 2005 revealed the Administration acknowledging Exxon's active involvement in determining its policy on climate change, as well as that of the Global Climate Coalition . . . an entity that hawk-eyed readers may recall from that long list on pages 228-9.

Ron Arnold (b1937), a logging consultant, onetime director of the Reverend Sun Myung Moon's Unification Church front group the American Freedom Coalition, and one of the primary scourges of environmental protection in the US, spelled out the tactic of industries funding and if necessary creating *faux* activist groups in 1980 in *Logging Management Magazine*:

> Citizen activist groups, allied to the forest industry, are vital to our future survival. They can speak for us in the public interest where we ourselves cannot. They are not limited by liability, contract law, or ethical codes . . . [I]ndustry must come to support citizen activist groups, providing funds, materials, transportation, and most of all, hard facts.

All of which might seem reasonable enough, except that the "hard facts" on offer are usually anything but, instead being deliberate misrepresentation – or, to use Milloy's Orwellian jargon, "sound science". Arnold himself perverted the concept of "wise use" (of forests), originally developed by Gifford Pinchot in *A Primer of Forestry* (1903) as the need to balance the demands of nature conservation against commercial interests, to mean the wholesale exploitation of wild areas and to hell with the consequences; we (i.e., someone else, later) could more easily solve problems created by the destruction than we (i.e., industry) could forgo the immediate profits.

The timber and mining industries gleefully funded the "wise users" while their campaigns against environmentalists included death threats and arson. These industries cannot have been unaware that the groups they funded were associating themselves with some extremely sordid allies, including the right-wing militias; a crossover organization was the innocuously named National Federal Lands Conference, which sought to overturn at county level the federally guaranteed protections of wild lands and consequently, because of its anti-federal venom, attracted militant right-wing extremists.* Only when two men associated with the fringes of the movement, Timothy McVeigh and Terry Nichols, driven by the propaganda of the "wise users" and others to the insane belief that anything was permissible in the fight against "government meddling", in 1995 bombed the Murrah Federal Building, Oklahoma City, killing scores, did the logging industry tiptoe quietly away from the "movement" it had founded. More recently the wise users have formed alliances with far-right Christian fundamentalist groups (the ideological link being that many of these groups believe mankind is divinely ordained to exploit all natural resources to the hilt), so one can probably expect another outrage like the Oklahoma City bombing in due course. Matters have not been helped by the efforts of the likes of Pat Robertson (b1930) and his Christian Coalition to replace communists with environmentalists

---

* Antisemitism is, for some reason, another characteristic of the NFLC – and of course another attractant for the right-wing militias.

as the boogeymen his faithful flock should irrationally fear and detest.

There are still plenty of industry-funded "astroturf" bodies around in the US today – so named because they're fake grass-roots organizations. Their industry funding is frequently laundered through other apparent non-profit groups, which in turn take "service fees"; an example is Americans for Tax Reform, as used for laundering by, among others, Jack Abramoff (b1958). Among astroturf organizations that seemingly concern science/technology issues, many of which, following governmental examples such as the Clear Skies Initiative, deploy the Orwellian technique of having titles that indicate exactly the opposite of their real purpose, we can note:

• Americans for Technology Leadership: Partly if not entirely funded by Microsoft, this covertly organized a letter-writing campaign in support of the company when it was facing anti-trust litigation. Supporters of Microsoft were given pre-printed, "personalized" letters to send. Quite how many of these letters were "sent by" individuals who had no idea of what was going on is difficult to ascertain; but the Attorney General of Utah discovered that at least two of the "senders" were dead at the time.

• American Taxpayers Alliance: Funded entirely by Duke Power and Reliance Energy, two of the power companies involved in the California energy crisis of 2000–2001, this organization played a major role in bringing about the down-fall of California Governor Gray Davis (b1942), who was trying to curtail the power companies' profiteering and market manipulation.

• Save Our Species Alliance: Seemingly funded by the logging and paper industries, this organization was behind the 2005 bill sponsored by Congressman Richard Pombo (b1961) that would have effectively gutted the Endangered Species Act by making it prohibitively expensive for the US Government to protect habitats from logging.

• Project Protect: Allied to the American Forest Resource Council, an industry group dedicated to supporting logging interests, Project Protect purported to represent ordinary citizens incensed by the resistance to President Bush's "Healthy Forests" initiative, eventually realized, thanks in large part to

Project Protect, as the Healthy Forests Restoration Act – an Act that was another Richard Pombo favourite since it "restored" forests by permitting more of them to be destroyed by logging companies.

• Citizens for Better Medicare: A front organization for the pharmaceutical industry, dedicated to stopping the US Government from bringing down the cost of prescription drugs through the creation of a prescription drug benefit. The organization spent about $65 million, largely in TV advertising, on a scare campaign.

• United Seniors Association: Although this claims 1.5 million individual members, its funding seems to come almost entirely from the trade association Pharmaceutical Research Manufacturers of America. Like Citizens for Better Medicare, this focuses on prescription-drug legislation.

• 21st Century Energy Project: Funded covertly by Enron (through Americans for Tax Reform) and Daimler–Chrysler (through the rather similar Citizens for a Sound Economy), this was responsible for an attack campaign against those who opposed industry-favouring energy legislation promoted by President Bush.

• Citizens for Asbestos Reform: Sharing an address with the American Insurance Association – its primary funder – this organization lobbied in support of the 2003 Asbestos Injury Resolution Act, which reduced liabilities for injuries and deaths caused by asbestos.

How effective are the astroturf organizations in disseminating false science among the minds of the public? Really quite effective, if we look at the response to the publication in early 2007 of the latest report from the UN-sponsored Intergovernmental Panel on Climate Change. Compiled from the research of hundreds of climate scientists worldwide, this set out definitively what had already been clear since the late 1990s: that the activities of humankind are accelerating global warming, and that the point of no return, after which catastrophic climate change will be inevitable, is extremely imminent. Treated as if it were a serious contrary argument was the ludicrous industry-generated "talking point" that all those hundreds of scientists were simply *saying* climate change existed because they were worried about losing their jobs!

Above: A front page from *Das Schwarze Korps*, the SS journal
Backdrop: Heinrich Himmler

CHAPTER 6

THE POLITICAL CORRUPTION OF SCIENCE

---◄◙►---

INTRODUCTORY

SINCE THE DAWN of the 20th century there has been an upsurge in the ideological corruption of science by those in political power. The three classic cases over the past century are Hitler's Germany, Stalin's USSR, and George W. Bush's America. The third of these is treated in slightly more detail here not just because it is current but because its ideological spread has arguably been on the widest scale, affecting not one or two sciences but many. It is also, because of its assault on climate science, potentially the most globally dangerous in human history. This should not be read as reflective of any political attitude the author might have for or against other activities of the Bush Administration; the deliberate governmental corruption of a nation's science is of such parlous importance that it transcends all political allegiances or antipathies.

Allied to governmental corruption and/or suppression of science, there's (as noted before) a quaint perversion of democratic thought which seems to believe that scientific truth can be determined by populist vote – or at least by opinion polls. The playwright Nathaniel Lee (c1649–1692) supposedly gibed, "They called me mad, and I called them mad, and, damn them, they outvoted me." In terms of sanity, "their" vote may have been of note, in that sanity is largely defined as conformity to society's norm; but in other areas of science a voting exercise – most especially if most of the voters are ignorant of or ill-informed about the issue at hand – is meaningless. Reality is not affected by human preferences, and most certainly does not obey human wishes: reality simply *is*.

The political corruption of science goes back a long way. The Roman physician Galen (129–*c*200) is perhaps the single most important figure in the history of medicine: among his many achievements was the recognition that medical treatment should be based upon experiment and observation rather than abstract theorizing. Yet his science was irremediably corrupted by the political climate of Rome, which forbade human dissection. Galen therefore had to make do with dissecting animals, under the assumption that human anatomical features would be of similar nature to those of sheep, pigs, etc., even if different in detail and in disposition around the body. Among his false conclusions were that the human heart had two chambers (in reality it has four) and that blood was manufactured by the liver. Because of Galen's (deserved) pre-eminence, his misconceptions corrupted medicine for centuries.

This leads us to a significant point. The corruption of medical science consequent upon the stance of Galen's contemporary Roman politicians does not imply that those politicians were themselves corrupt in banning human dissection – in fact, they were displaying admirable integrity in insisting that scientists should act according to the moral standards of the day. Although it's certainly true that the vast majority of political corruptions of science are the products of bogus ideology and/or sleaze, this is not inevitably the case: honest politicians can make honest decisions without realizing the disastrous effect these can have on science. In the current furores over embryonic stem cell research, for example, while undoubtedly the majority of politicians stumping in favour of a ban are doing so for corrupt reasons, pandering for votes, it's equally certain that a few political supporters of such a ban are acting in the honest belief that on occasion morality is more important than scientific advance. They may be wrong, and they may arguably be stupid, but they are not necessarily corrupt.

HITLER'S GERMANY

――――――⋖⊛⋗――――――

If science cannot do without Jews, then we will have to
do without science for a few years.
  Adolf Hitler, 1933

AFTER WWII, when it became possible to evaluate the
damage done to the sciences in Germany by the corruption of them fostered by the Nazi regime, the surprising
conclusion was, at least in physics, chemistry and mathematics, and at least in the shorter term: not much. To be
sure, the vast majority of the best scientists in these fields
had departed during the 1930s for more tolerant climes
abroad, while a number of promising up-and-comers had
incurred the wrath of the Nazis and been killed,* but the
German scientific tradition had been a strong one: in
essence, there were still plenty of good German physicists,
chemists and mathematicians to go round, and for the most
part they had been permitted by the authorities to carry on
their science as usual, without interference, so long as they
kept quiet about it. There were temporary victories for the
lunatic fringe, but by the end of the war these had largely
been annulled. The scientific craziness that was supported
by the Reich was essentially for public consumption; in
private the Reich realized how necessary good science was
for its own glorious 1000-year survival. Even as powerful a
man as Heinrich Himmler (1900–1945), whose espousal of
the pseudosciences was profound, was sidetracked into his
own "scientific" organization, the Ahnenerbe.

There was one area, though, where mathematics and
the physical sciences suffered grievously under the Nazis,
and it was to be some while before they could recover. This
was in education – and the lesson of what happened is one
that is valid today, when schools elsewhere are under

---

* A number, too, died fighting for their country.

constant pressure to include "alternative viewpoints" in their science curricula. While German physicists and mathematicians might have been working with or exploring Relativity and quantum mechanics in their research, when it came to teaching students they had to be far more circumspect; furthermore, many professors and lecturers were dismissed and either replaced by Party hacks or not replaced at all. This meant that in the longer term German science suffered considerably: a whole generation of university students, and a further generation of schoolchildren, were woefully miseducated in the physical sciences, being taught a version of science that contorted itself ludicrously in order to omit or disparage the keystones of 20th-century science. Even today the effects of this ideological corruption can be felt: while Germany's standing in the scientific world is by no means shoddy, it is still far, far below the heights it reached in the decades before its assault by Nazism.

The biological sciences fared far worse in the shorter term because of their complete corruption by racist ideology. Medicine did not escape either: while some of the medicine the Nazis promoted was surprisingly progressive – for example, Hitler had a phobia about cancer, so cancer research and prevention were heavily promoted – the notions of medical pseudoscientists and ideologically sound quacks were presented as if they had validity. Further, the horrific experimentation on live subjects carried out at concentration camps left, like the use of slave labour in the technological industries, a stain on German science that was near-indelible. And, again, the longer-term effects of miseducation in the biological sciences did grievous harm to their practice in the nation for fully a generation. As to the effects of brainwashing young people that they should see the world, science included, only through the lens of fascist ideology? Who knows how much damage that has done.

In considering the corruption of science under the Nazi regime in Germany, it's useful first to consider the different behaviours of the German scientists who lived – or in some cases did not live – through that regime. There were some who, for reasons of ideological belief or straightforward opportunism, actively cooperated with the regime: examples are physicists Philipp Lenard (1862–1947), Johannes

Above: Philipp Lenard

Left: Johannes Stark

Stark (1874–1957) and Ernst Pascual Jordan (1902–1980), zoologists Eugen Fischer (1874–1967) and Otmar von Verschuer (1896–1969), anthropologist/zoologist Konrad Lorenz (1903–1989) and physician Josef Mengele (1911–1979). There were some who did not support Nazism *per se* but who nonetheless contributed, enthusiastically or tepidly, to the German war effort either through a misplaced sense of patriotic duty or, again, through opportunism, in that the regime was offering extensive funding and a reality of scientists' lives is that they must pursue funding if they're to continue their research: examples are Werner Heisenberg (1901–1976), Otto Hahn (1879–1968) and Wernher von Braun (1912–1977). A few remained in Germany but offered covert – or in a few brave instances overt – resistance to the regime and its efforts: physicist Max Planck (1858–1947) was one, although his resistance was more in thought than in deed (to be fair, he was in his 80s when the war began). There were those who were persecuted, interned, starved and tortured by the regime: by far the best known example of the few in this category who survived is the chemist Primo Levi (1919–1987), who somehow lived through the hell of Auschwitz to tell, most movingly, the tale. And finally there were the many, many scientists who fled from the regime; examples are far too numerous to list, but we can mention as of especial note Albert Einstein (1879–1955), Lise Meitner (1878–1968), Leo Szilard (1898–1964) and even Sigmund Freud (1856–1939). Many of the physicists who fled in due course became participants in the Manhattan Project.

Leaving Mengele aside for the moment – not to mention the numerous anthropologists who strove vainly to find any scientific rationale whatsoever for the Nazis' dream of Aryan racial purity – the person whose science was most significantly, and in any other context hilariously, corrupted by the Nazi ideology was the mathematician Ernst Pascual Jordan. At the time all the rage in the physics world was quantum mechanics, a theory that had been pioneered and elucidated largely by German scientists. However, the great clarion cry of the Nazi regime was that theorization was airy-fairy "Jewish science" and therefore to be rejected; instead physicists should be concentrating on *Die Deutsche Physik* (see page 245), where the emphasis was on experimentation and the focus on the tangible. Jordan, a fervent Nazi, set himself the difficult, indeed self-contradictory, task of showing that quantum mechanics was actually a scientific analogue of Nazism, and that the two – the physical theory and the political ideology – therefore supported each other.* Just as the new physics was in due course going to explain all areas of reality and tie them together as merely different aspects of a unified whole – and Jordan was not afraid to include such items as religious experience and

* Of course, many in the USSR maintained that Darwinism was a capitalist theory; while ludicrous, this was not *quite* as ludicrous as Jordan's thesis, in that raw capitalism, with its notion that survival of the fittest is a desirable, even ethical code upon which to base human interactions, can easily be regarded as the social equivalent of Darwinism. But both stances – just like the rabid capitalist's citation of Darwinism as a justification of capitalism's worst excesses – display an astonishing degree of ignorance as to what a scientific theory actually is.

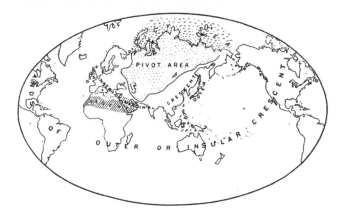

Mackinder's geopolitical map of the world

clairvoyance on his list along with cosmology and thermo-
dynamics – so Nazism was not merely a political ideology
but a unifying explanation of human existence and a
prescription for future human social evolution. In short,
both were Grand Universal Theories.

One is reminded of Richard Dawkins's remark concerning the proliferation of books linking quantum mechanics
and cosmology to Eastern mysticism: their thesis seems to
be that quantum mechanics is hard to understand and
Eastern mysticism is hard to understand, so the two must
really be the same thing. Jordan's logic in trying to equate
physics and Nazism seems to have followed this principle.
There were too many other German scientists willing to
follow his example.

GEOPOLITICS

The idea of geopolitics was initially the brainchild of the UK
diplomat and geographer Halford Mackinder (1861–1947).
In his paper "The Geographical Pivot of History" (1904)
and elsewhere he set out to show how much of humankind's
political history had been – and by extension would in
future be – determined by geography, with particular focus

on what he would later call the World Island, the great
Eurasian continent. Geographical location governed
resources; relative position with respect to landmasses and
oceans governed the accessibility and defensibility of a
culture; taken together, these factors could confer a significant advantage to luckier cultures in war, and hence in
those cultures' level of dominance in world politics. Any
culture that controlled the interior – which Mackinder later
called the Heartland – of the World Island was likely to
control the world. Much later he was to summarize these
ideas in a triplet:

> Who rules East Europe commands the Heartland.
> Who rules the Heartland commands the World Island.
> Who rules the World Island commands the world.

Mackinder's boundaries for the Heartland were approximately those the USSR would come to have after WWII. A
corollary of his main thesis was that sea power, regarded as
paramount by the militarists of the day, would inevitably
wane in importance as transportation and communications
surged ahead within the Heartland (which in Mackinder's
day meant the coming spread of Eurasian railways). This
was not to say naval power would become insignificant; one
passage in Mackinder's 1904 paper chills today because of
its seeming prescience:

> The oversetting of the balance of power in favour of the pivot
> state, resulting in its expansion over the marginal lands of
> Euro-Asia, would permit the use of vast continental resources
> for fleet-building, and the empire of the world would then be
> in sight. This might happen if Germany were to ally herself
> with Russia.

Though Mackinder's ideas seemed plausible, for some
decades they went largely ignored in the English-speaking
world. However, they attracted the attention of the German
retired general Karl Haushofer (1869–1946) and his group,
who in the 1920s and 1930s merged them with ideas of
their own to produce the concept of *Geopolitik*. The timing
could not have been direr: just while the Nazi Party was
rising to ascendancy, along came a quasi-scientific theory
that not only spelled out the geographical "entitlement" of

the supposed descendants of the mighty Aryan races but also offered a blueprint for how those descendants could regain their "heritage". Hence was born the notion of *Lebensraum*, usually translated as "living space"; what the bald translation omits is that the concept had a considerable underpinning of dubious science, both bad extant evolutionary science, which considered that each species (and by extension each of the human races) had its own natural, pre-ordained territory, and Mackinder's appealing but empirically dodgy hypothesis.*

Mackinder was not entirely blameless in the elevation of his concepts to military-political dogma: in his 1919 book *Democratic Ideals and Reality* he emphatically proposed that, while good people might hope for democracy to conquer the world through conquering people's hearts everywhere, the truth was that the only determinant of political and economic clout was the increase of territory through war. While he himself claimed to recognize this fact only to prescribe ways of avoiding or averting such conflict, in reality he provided a blueprint not so much for the peacemakers as for the Nazi warmongers. For Germany to regain its *Lebensraum* it had to form a geographical union with Russia, through either conquest or alliance, and so the necessary prerequisite was to eliminate the belt of buffer states the post-WWI negotiators had created between the two: hence, as a first step along this route, the invasion of Poland.

DIE DEUTSCHE PHYSIK

The masterminds of *Die Deutsche Physik* – "German physics", which rejected abstract theorization as "Jewish" and thereby a false representation of reality – were Philipp Lenard and

---

* Mackinder regarded Australasia, Japan, the Americas and all but the northernmost part of Africa as peripheral in both geography and, as a consequence, in potential for global influence. In light of the world's current balance of power, this seems somewhat of a death blow to his hypothesis. To be fair, however, we have no more idea in 2007 what the balance of power will look like a century hence than Mackinder did of today's political equation when he was writing in 1904.

Johannes Stark. Although one can to an extent understand the direction they were coming from, in that the crux of the Scientific Method is not to theorize in a vacuum but to construct tentative hypotheses based on observation and experiment, then test them through further observation and experiment, it is obvious that without the work of the theoreticians science would be in a sorry and backward state; it is equally obvious that to make a racial identification, as if the ability to theorize were hereditary, is nonsensical.* Such obviousness was lost on Lenard and Stark, which is all the more surprising in that they were both deserving recipients of the Nobel Physics Prize, Lenard in 1905 and Stark in 1919. They serve as a graphic reminder of the fact that even the most brilliant scientists can be fools outside their own narrow disciplines.

Lenard was a profound hater. He hated his colleagues, he hated the English, he hated just about anyone who got in his way, and above all he came to hate the Jews. Oddly, he at first hailed Relativity; it was only later, on account of Einstein's Jewishness, that Lenard became a leader of the motley ratpack who denounced Relativity as a sham and attempted, with little success, to install *Die Deutsche Physik* in its place. The name was enshrined in Lenard's four-volume magnum opus *Deutsche Physik* (1936), in which he maintained that genuine physics must be both racially conformable to the physics community (*blutmässig*) and intuitive (*anschaulich*), by which latter term he seems to have meant commonsensical (unlike the abstractions of Relativity and quantum mechanics) and based on direct experimental observation: those mysteries of nature unamenable to such observation would, and should, remain forever beyond our

* The inclination and ability towards abstract thought would seem to rely upon intelligence, certainly, but more significantly upon youthful environment; i.e., education, both formal and in terms of the domestic climate. Since in any nation there are observable disparities between the attitudes of different communities towards education in this broad sense, there are bound to be equivalent disparities in the percentages of "theorists" emerging from those communities. The high ratio of Jews among German scientists reflected such a cultural difference, which was in turn encouraged by the prevalent antisemitism: nobody was inviting young Abraham to the bierkeller, so he stayed at home and studied.

ken. True physics should be *unimaginative*, which was why the Jewish influence on physics was so toxic: Jews were imaginative, and thus in the thrall of theoretical speculation. Any Jew who was good at *Die Deutsche Physik* was so only because he (the notion of it being a she was far beyond Lenard's mental horizon) must have some Aryan blood in him – as for example the part-Jewish Heinrich Hertz (1857–1894), under whom Lenard had studied; Hertz's Aryan blood was responsible for the great experimental feat of discovering radio waves, but his Jewish blood betrayed him later in life when he ventured into the cesspit of theory. Those theoretical physicists who were unquestionably Aryan, like Heisenberg, were "white Jews".

Stark was in essence a disciple of Lenard, so far as their bureaucratic activities under the Reich were concerned; a dozen years younger, he was more able to put into practice the ideals of the older man. On the subject of the "white Jews" he expounded in 1937 in the SS journal *Das Schwarze Korps*:

> And if the bearer of this spirit is not a Jew but a German he is all the more to be combated than the racial Jew, who cannot conceal the fount of his spirit. For such carriers of infection the voice of the people has coined the description of "White Jew", which is particularly apt because it broadens the conception of the Jew beyond the merely racial. The Jewish spirit is most clearly discernible in physics, where it has brought forth its most "significant" representative in Einstein.*

In the spring of 1933, a few months after having been sworn in as Chancellor, Hitler announced that all Jews working for the civil service should be purged from their posts. At the time, the structure of the scientific establishment in Germany was such that all academic scientists were civil servants, so the immediate damage to Germany's position as a leading scientific nation was enormous. Many of the nation's physicists, fully aware of this, raised loud or timid voices in protest. (It was in response to special pleading by

---

* As cited in translation in Walter Gratzer's invaluable *The Undergrowth of Science* (2000).

Max Planck that Hitler made the remark at the head of this subchapter.) Others, opportunistically, saw a chance for promotion into the vacated posts, and kept their peace. And yet others, Lenard and Stark at their forefront, welcomed the "cleansing" of German science.

Einstein, who fled, was a particular target for the excoriations of this evil pair. Heisenberg, who stayed, came in for his fair share of abuse as well, and over him Lenard and Stark had some actual power: they could block his appointments and they could make sure Himmler's SS targeted him for special investigation. Although there could be little future in attacking Heisenberg's Aryan credentials, which were impeccable, through his history and widely recognized abilities as a theoretician they could get round that trivial little difficulty: he was the archetypal "white Jew". Another epithet they used against him was that he was the "Ossietzky of physics".[*] Among his stated crimes were that he defended Relativity and had worked collegially with Jews. Heisenberg's future would have looked grim had it not been for the chance that his mother knew Himmler's mother: Mrs Himmler promised to have a word with her boy. For about a year Himmler dithered, but finally Heisenberg's name was cleared of all "charges" and Lenard and Stark were told to shut up. Hilariously, Himmler, in conveying this decision to his sidekick Reinhard Heydrich (1904–1942), expressed the hope that Heisenberg might be able to help in the elucidation of the World Ice Theory (see below).

Also especially singled out for abuse was the University of Göttingen, supposedly a hotbed of the Jewish conspiracy to poison Aryan science. In fact, there were a lot of Jewish scientists there, especially at the Mathematical Institute,

---

[*] Carl von Ossietzky (1888–1938) was a journalist and pacifist, co-founder in 1922 of *Nie Wieder Krieg* ("No More War") and editor from 1927 of *Die Weltbühne*, in which weekly he denounced Germany's secret rearmament. He was sentenced to prison in 1931 for treason, but the sentence was commuted and he returned to his editorship. On Hitler's accession to power Ossietzky was sent to the Papenburg concentration camp. There he received the 1935 Nobel Peace Prize but eventually died of tuberculosis and maltreatment.

where Christian Felix Klein
(1849–1925) – of Klein Bottle
fame – was one of the stars.*
Whether Klein had actively
encouraged the promotion of
Jews here and at other German
universities as a matter of pro-
semitic bias, or simply because he
sought the finest mathematicians
and many of these happened to
be Jewish, is something we don't
know, but he was certainly
attacked posthumously by the
Nazis for the former. It seems

Klein

likely both motives played their part, because from the
1890s onward he put forward the hypothesis that different
races conformed to different mathematical aptitudes:

> Finally, it must be said that the degree of exactness of the intu-
> ition of space may be different in different individuals,
> perhaps even in different races. It would seem as if a strong
> racial space intuition were an attribute . . . of the Teutonic
> race, while the critical, more purely logical sense is more fully
> developed in the Latin and Hebrew races.

Klein's hypothesis was intended to support what we'd now
call multiculturalism. Unfortunately, he used for this a term
meaning "racial infiltration", and it was for his approval of
"racial infiltration" that the Nazis attacked him. Further, his
own mathematics marked him out as a "white Jew" *par excel-
lence*. Even so, his racial stereotyping was something they
could use as yet another cudgel with which to beat Jewish
science, and no one took it up more eagerly than Klein's
ex-student Ludwig Bieberbach (1886–1982), by now a

---

* At one stage when Klein was under surgery he was given a blood trans-
fusion by his colleague Richard Courant (1888–1972), a Jew. Thereafter
it was quipped that even Klein, a rare Aryan mathematician at
Göttingen, had Jewish blood in his veins. Courant had been the founder
of the Mathematical Institute, in 1920. Much later, in 1933, he fled
Germany, first for the UK and then for the US, where he pursued a
mathematical career of distinction for the rest of his life.

Ludvig Bieberbach

professor at the University of Berlin.

The tragedy of Nazism is almost always seen as the tragedy of the victims of the Holocaust, and very rightly so – who could say otherwise in the face of millions of deaths and untold torments? Yet this is to ignore the fact that Nazism destroyed also so many of its staunch supporters. One of these was Bieberbach. Looking at his record before he became infected in the early 1930s by the Nazi virus (he was, curiously, if anything a liberal early on), one might have predicted he'd emerge as one of the 20th century's heroes of mathematics. Instead he's remembered, if at all, as a maimer of German mathematics (replacing it with *Deutsche Mathematik*, of course) and indirect enabler of the Holocaust.*

Bieberbach came to believe mathematics should be intuitive and commonsensical – *anschaulich*, in other words, just like Lenard's *Deutsche Physik*. This brought him into direct confrontation with the modernist school of mathematics spearheaded by David Hilbert (1862–1943), who had succeeded Klein at Göttingen and is today regarded as among the most important figures in mathematics history. It helped Bieberbach's campaign that Hilbert was profoundly anti-Nazi: Hilbert had bitterly resisted the dismissal of Jews from Göttingen during the general purge of 1933 and, crime of all crimes, had, many years earlier (before WWI), been influential in the University of Göttingen's refusal of a chair to Johannes Stark; Stark blamed this rebuff on a "Jewish and pro-semitic circle" at

---

* The full story of Bieberbach's corruption of German mathematics is outwith the scope of this book: what follows is the briefest sketch. For an excellent account see Sanford L. Segal's *Mathematics Under the Nazis* (2003), in particular its near-book-length Chapter 7, "Ludwig Bieberbach and 'Deutsche Mathematik'".

the university. Hilbert and
his school were, in other
words, easy targets at a
time when political ideology
transcended science. As one
example of Bieberbach's
folly, he excoriated at
Göttingen the (Jewish) mathematician Edmund Landau
(1877–1938) for the crime of
teaching the expression of pi
as the sum of an infinite
series.*

David Hilbert

Bieberbach continued
his rabble-rousing efforts
through the 1930s, and had a
certain success among the young – students and junior
academics. Internationally, however, and among established
mathematicians, his attempts to redefine German mathematics were regarded with anything from horror to
incredulity to ridicule, and they suffered from the fact that
the Nazi hierarchy weren't much interested, probably
because, while the Nazi leaders thought they could understand physics, they knew maths was beyond their grasp.
Bieberbach made a number of political moves within
various mathematical organizations in an effort to impose
his views on the mathematical community from a position of
power if he could not do so by persuasion or rhetoric, but
these too eventually came to nothing.

As a consolation prize Bieberbach was given the editorship of and funding for a new mathematical journal,
Deutsche Mathematik, whose first issue appeared in January
1936. This contained genuine mathematical papers inter-
spersed among political exhortations and philosophical
ramblings about racial stereotyping along the lines of ideas,
which Bieberbach had espoused, promulgated earlier by
the psychologist Erich Jaensch (1883–1940): there were two

---

* The irrational nature of the number pi has affronted countless fools
over the centuries. For example, in 1897 the Indiana State Legislature
came within two votes of declaring the legal value of pi within the state
to be 3.2.

types of human mind, racially determined, the J-type, which conformed to the *anschaulich* ideal and was thus laudably Aryan, and the S-type, possessed by inferior folks like Jews, most French mathematicians, and in general anyone disliked on ideological grounds. Despite or because of such twaddle, the journal was very successful – perhaps also because it was deliberately kept much cheaper than other scientific journals, and thus more accessible to student budgets.

After the end of the war, Bieberbach continued to maintain his views on racially determined mentalities while at the same time projecting wide-eyed innocence about the concentration camps: so far as he was concerned, expelling people from their jobs and homes was merely benign racial purification, and he'd had no idea what happened to them thereafter, why should he? In fact, it wasn't until the mid-1950s that he even acknowledged the existence of the camps. It seems he was one of the earliest of all Holocaust-deniers, and all the more remarkable in that he cannot but have known what was going on all around him.

The failure to topple Heisenberg more or less marked the end of the *Deutsche Physik* movement: Lenard was growing progressively more obviously dotty while the abrasive Stark had rubbed too many of the Nazi hierarchy up the wrong way. Still, the Heisenberg episode was to be repeated in its essentials with other physicists all through the period of the Reich. Only in the imaginings of madmen could such absurdity be regarded as a sensible way to direct the course of science, yet such was the corruptive effect of the Nazi worldview that it seemed commonplace. Certainly Heisenberg must, incredibly, have thought this way of conducting affairs at the very least tolerable, because he decided still to stay on and practise science in his native land, in due course leading the Reich's efforts to develop an atom bomb.

Himmler's power to direct the scientific establishment, a power that varied during the years of the Reich but at times approached the absolute, was something the man fought hard to gain, despite the fact that he lacked scientific qualifications. (So did Hitler, of course, but this didn't stop crazies like Lenard and Stark hailing the Fuehrer as a scientific genius.) This was all the more disturbing in that Himmler was an aficionado of the pseudosciences, not just the pseudosciences of racial purity and Aryan superiority but also astrology and the occult.

Himmler was fascinated by science and at the same time a complete scientific illiterate. Either to limit his meddling with the real science underway in Germany or so that the Reich might explore arenas of science the academics were inexplicably reluctant to pursue, he had been permitted in 1935 to set up his own research institute, the Ahnenerbe. The name means roughly "ancestral heritage", and indeed a large part of the "research" done there was into tracing the glorious Aryan lineage of the Teutonic peoples. In Sam Goudsmit's book *Alsos* (1947), cited in Gratzer's *The Undergrowth of Science*, is reproduced this 1944 letter from Himmler:

> In future weather researches, which we expect to carry out after the war by systematic organization of an immense number of observations, I request you to take note of the following: "The roots, or onions, of the meadow saffron are located at depths that vary from year to year. The deeper they are, the more severe the winter will be; the nearer they are to the surface, the milder the winter." This fact was called to my attention by the Führer.

Himmler was ever ready to accept the ramblings of amateurs like himself over the hypotheses of trained scientists, and there was really nothing any sane person could do to stop him: dissent was equated with treason, and so arguing with Himmler was the fast track to a death camp.

This attitude of Himmler's spread over into technology, specifically weapons technology, where researchers had no choice but to head down one pseudoscientific alley after another at Himmler's whim. One device they were forced to attempt relied for its functioning on the notion that the atmosphere contains an "insulating material" vaguely analogous to the luminiferous aether thought for some decades to be the omnipresent medium through which electromagnetic waves travel, but in this instance a property of air rather than vacuum. Without the insulating material, so a pseudoscientist had persuaded Himmler, electrical equipment wouldn't work: it'd short-circuit. The aim of the hypothetical device that could help win the war for the Reich was to be able, from a distance, to strip away the insulating material from the air around the enemy's electronics.* It sounds like something the mad scientist develops in a James Bond movie, and every bit as plausible. Another Himmlerian brainstorm triggered research into the building of a 1000-ton tank, which might indeed have been near-invincible to enemy fire but would have had an interesting experience whenever it encountered a structural weakness in the ground below. Then there was the scheme to harvest fuel alcohol from the chimneys of German bakeries . . .

One bizarre notion rife among the Nazi leadership was that disease could be caused by "earth rays". Precisely what these were is hard to establish, but they appear to have been related in some way to magnetism: Rudolf Hess (1894–1987) hung magnets around his bed so that, while he slept, these would draw out of his body any harmful earth rays it might have absorbed. One characteristic we know the earth rays did have was that they were detectable by dowsing; dowsers were often hired to check out Nazi public buildings for any lethal emanations that might be lurking.

Had it not been that so many of the Reich's most able researchers were wasting their time on madcap programs there's every possibility Hitler might indeed have won WWII. Just think of the situation had the same resources been poured into the German quest for the atomic bomb.†

Another piece of pseudoscience embraced by at least some of the leaders of the Reich's scientific establishment was the World Ice Theory or *Welteislehre*, an invention of the Austrian mining engineer Hanns Hörbiger (1860–1931), who in 1913 published his speculations in a vast tome called *Glazial-Kosmogonie*. Here they can only be summarized.

* There's a germ of truth in this, of course. Air itself isn't a very good conductor of electricity, although high voltages can persuade it to conduct – hence lightning.

† The grapevine has it that the team dedicated to this quest, led by Heisenberg, was "close" by the end of hostilities. In fact, it had barely got beyond the basics. The German scientists, by then in captivity in England, were as stunned as anyone else by the news of the Hiroshima detonation.

What we think of as stars are mere chunks of ice. The only star (in our sense of the word) is the sun, which at the beginning of things was some 30 million times larger than it is now. Everything else, the earth included, orbits the sun; the earth is the only other celestial object not to be a chunk of ice or at least entirely ice-covered. Space is not a vacuum but filled with the aether, which is in actuality rarefied hydrogen; as celestial objects move through this, friction causes their orbits around the sun to decay so that they spiral inwards until finally

Hanns Hörbiger

falling into the sun, creating sunspots. When stars do this, their ice is vaporized and part of it explodes back into space; some of the ice reaches the earth refrozen as tiny crystals, which we see as high-altitude clouds. They melt and reach the ground as rain. Bigger crystals, which we see as meteors, explode on entering the atmosphere, and their fragments fall as hail.

Mars is covered in an ocean of ice or water some 400km deep, and will one day become a moon of the earth – unless it misses us and instead plunges straight into the sun. In fact, the earth itself – like all the other planets – will in the end fall into the sun. This is not to imply the end of the solar system: there is, apparently, an infinite number of planets beyond Pluto, and these are slowly approaching the sun as if on some cosmic conveyor-belt. Thus, even though the planets we know will die a fiery death, others will take their places. Long before our world plummets to its doom, however, the moon, which is spiralling in towards us, will have collided with it. This will not leave our devastated planet moonless: as with the planets and the sun, there are replacement moons queuing up. Indeed, this is not our world's first moon: the most recent of its several predecessors fell to earth just a few thousand years ago, a catastrophe we know from various myths and legends, notably that of the Flood. The capture thereafter of the next moon, which is the one we see in our skies, caused further upheavals, such as an outbreak of poleshift* and the sinking of Atlantis, which continent was itself a relic of the previous fallen moon.

Hörbiger claimed that the World Ice Theory came to him fully formed in a dream, and reading the above we might not be surprised: what, one wonders, had he been smoking? Even so, the theory became inextricably involved with the rising tide of German Nazism. This cosmology was as different from the "Jewish" one as any loyal Aryan could hope for. That it came from an Austrian amateur scientist was an added advantage: after all, was not Hitler himself an Austrian amateur scientist? Further, a reason for the Aryans' natural superiority to all the rest of mankind was obviously because their ancestors had been toughened up in the chill of their northern habitat during the last glacial age. Ice again! Surely it couldn't be coincidence! Indeed, to reject the World Ice Theory would be to reject the very notion of Aryan superiority.

Hitler and especially Himmler loved this.† It seems what attracted Himmler to the theory was the bit about the ice crystals falling into the atmosphere. Accurate weather forecasting is a more than useful ally when a country is waging war. If the enemy, through foolishly ignoring Hörbiger's uncovering of the true nature of reality, were unaware of where weather actually came from – outer space, of course – then their meteorologists would be at a marked disadvantage against German ones. Himmler set up an Institute of Meteorology whose scope was immediately

---

* The idea of poleshift is that the direction of the earth's rotational axis periodically flips, with widespread disaster as a consequence. There would in truth be widespread disaster – far more, indeed, than any of the poleshift theorists seem capable of imagining – if poleshift were feasible.

† The *Welteislehre* is of course a theory, and thus logically the Nazis should have put it into the "Jewish science" category. They didn't.

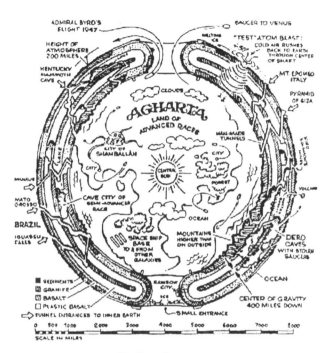

Teed's Hollow Earth

expanded to include all kinds of other disciplines in an effort to prove the *Welteislehre*. A surprising number of supposedly reputable scientists went along with this, through fear or otherwise. It was, however, too much for even the arch-bigot Philipp Lenard to stomach; his polemics against effort being wasted on such nonsense were sufficient to drive Himmler's researches, if not underground, then at least into the shadows, where they continued. And woe betide any lesser figures of the German scientific establishment than Lenard should they speak out against the theory!

The World Ice Theory was only one of the many crack-

pot items of pseudoscience that found their way to the
Ahnenerbe. Another was, at least for a while, the hollow-
earth theory of Cyrus Reed Teed (1839–1908), a US religious leader and crank; where Teed's theory differed from
others was that he believed we lived *inside* a hollow earth.
Some special "explanations" were required; for example, if
you look directly upward at midnight, why do you not see
the lights of the cities on the other side of the world? Teed
proposed that light could travel only so far around the
concave surface of the earth before, as it were, taking a
nosedive into the ground. Such a limitation did not apply to
light of "arcane" frequencies – such as infrared. And so the
Nazis conducted an experiment whereby they hoped to spy
on the manoeuvres of the British fleet. They pointed their
instruments up at an angle of about 45°, and, to their
surprise saw . . . clouds!

Teed's hypothesis did, however, have a more substantial
effect on the development of Nazi technology. In 1933 the
city council of Magdeburg decided it would test the hypothesis by asking the team of rocketry scientists then working in
Germany – including von Braun – to fire a few rockets
upwards to see if they'd land in the antipodes.
Unfortunately the Magdeburg councillors underfunded the
project, and so the experiments had to be curtailed – by
which time the best performance of the *Raketenflugplatz*'s
launches had been a horizontal flight of about 300m.
Nevertheless, these experiments paved the way for the V1
and V2.

THE ARYAN DREAM

During the Third Reich, Darwinian evolution was accepted
as a fact, even by Himmler. Himmler did draw the line,
though, at the notion of the Aryans having evolved, even
though the rest of humankind had obviously done so. By
contrast, the Aryans had started off as little seeds embedded
in Hörbiger's cosmic ice crystals, but had then for some
reason been released down onto the earth, complete
with a godlike understanding of things like electricity –
thunder and lightning were belated manifestations of a
fierce war machine they'd created during prehistory – to

conquer Atlantis, until then populated by Asian immigrants.

Himmler also had the theory that the Nordic races had long ago almost completely exterminated the peoples of southern regions, but that the survivors had managed to breed the numbers back up again and were present in the world today as Hottentots and Jews. His basis for this scientific conclusion was that many African women have large rear ends – as displayed in figurines from that continent – and he'd noticed the same about Jewish women. One of the anthropologists from the Ahnenerbe, Bruno Beger (1911–1998), was instructed to look into the matter. The farce turned to nightmare when Beger pursued his investigations in Auschwitz.

The term "Aryan" has valid technical meanings, although these are little used today precisely because of the Nazi abuse of the word. Technically, "Aryan" refers in linguistics to the ancestor of the Indo-Iranian language family, and to that family itself; in archaeology/anthropology it refers to a pre-Vedic people of whom little is known save that they represented an early stage in the mutual assimilation of the Indian and Iranian cultures. By the end of the 19th century, the word was often used as a distinction from the Semitic cultures; popularly, it became a synonym for "gentile". The appeal of the word to racists was that it derived from a Sanskrit word meaning "noble" or "elevated"; hence the emergent mythology of a noble race that had expanded from India across Iran and Europe and was the highest expression of humankind.

The theoretical roots of Nazi racism can probably be traced to the Social Darwinist ideas promoted notably by Ernst Haeckel (see page 21). Haeckel was Darwin's and evolution's most vocal and influential supporter in Germany, although Darwin seems to have adopted the attitude towards Haeckel that he, Darwin, should "sup wi' a lang spoon". Haeckel's countryman, the great cell biologist Rudolf Virchow (1821–1902),* was another to be

---

* Virchow believed disease was a product of cell malfunction. The Darwinian idea that cell mutation could on occasion be advantageous was thus anathema to him.

notably chary about Social Darwinism (and about Darwinism in general). Virchov saw only too clearly the murky swamps into which Social Darwinism could lead.

Social Darwinism adapted the notions of evolution by natural selection and applied them to human societies. The upside of this was that it offered a very optimistic, progressive philosophy concerning the future potential of the human species: change was not something to be resisted but to be embraced. The downside was, as Virchow predicted, that the temptation was there to start "improving" a particular human culture by weeding out its "unfit" members and by resisting the influx of "inferior blood": this was the slippery path of eugenics, down which people like Francis Galton had already slid and down which many others aside from the Nazis would also slide, especially in the US but also in parts of northern Europe (see pages 200ff). In Nazi Germany the notions were especially toxic because they were married with the myth of the noble Aryan race. The glorious future of Germany depended upon distilling its people down to their essential Aryanism.

This craziness was further spurred by the proselytizing of the UK germanophile Houston Stewart Chamberlain (1855–1927), who promoted the notion of *Rassenhygiene*, racial hygiene – a term not dissimilar from the more modern "ethnic cleansing". The mentally and physically disabled were obvious "impurities", as were homosexuals and communists, and it was perfectly obvious that gypsies, blacks and Jews weren't of Aryan stock. Really, the list of minorities that were stains on the Aryan bloodline could be extended as far as the whim took you, and in any direction. Notably, the consequence of such thinking is that the victims are no longer individual people, just items in a category, and this attitude was encouraged by the regime, which referred to the minorities as subhuman: it wasn't people who were being slaughtered, just animals.

Physicians such as Alfred Ploetz (1860–1940) – founder in 1904 of the Society for Racial Hygiene (he coined the term) – made their own loathsome contribution, observing that medical science was capable of prolonging the lives, and thereby increasing the breeding opportunities, of substandard humans. A further justification for the extermination of the sick and impaired came from the ideas of the surgeon Erwin Liek (1878–1935), author of the best-selling book *The Physician and his Mission* (1926). Nazi medicine in general sought a holistic approach – prevention and therapy should focus on the body and lifestyle as a whole, since all was interconnected, rather than on the malfunctions of individual organs or systems. In proportion, this approach can of course be useful, and the Reich's accent on lifestyle means of prevention had a positive influence on overall public health. When applied to exclusion, however, it clearly has dangers. Herbalism and homoeopathy became the norm. The ideas of Liek, as expressed through his book, took this further, reintroducing a variation on the old idea of the bodily humours. According to Gratzer's *The Undergrowth of Science* (2000),

> He argued, besides, that pain was a benign secretion of a disorder and an essential part of the healing mechanism: to suppress it impeded recovery. Endurance of pain was, moreover, a prime virtue and one with which the superior races were better provided. More, to be sick showed a lack of moral strength – was indeed immoral. . . . This provided Liek's followers, if not Liek himself, with a justification for annihilating the sick.

Between January 1940 and September 1942, the Nazis gassed 70,723 mental patients on the grounds that their lives "were not worth living".

The anthropologist Eugen Fischer was another significant voice; his speciality was interbreeding between members of different races, a reasonable enough topic for investigation except when twisted through a racist ideology; Fischer played a part in the mass sterilization of those children born of French soldiers and German women during the 1919 French occupation of the Rhineland, and later, with Nazism in the ascendant, began promoting the analogy that Germany's ethnic minorities, notably Jews, were cancerous tumours feeding on the otherwise healthy German body.

The botanist Ernst Lehmann (1888–1957) founded the journal *Der Biologe* (and the Association of German Biologists) before Hitler's ascent to power, which provided him with the perfect opportunity to promote his fascistic views under the guise of biological science. Most of the journal's articles promoted racist pseudoscience; the pseudoscientific theme was perpetuated through items disputing Darwinian evolution, or even developing it: the biologist Gerhard Heberer (1901–1973), for example, claimed that Nordic Man was a recent, more highly developed species to replace *Homo sapiens*. (It's not hard to see how this hypothesis could be adopted by those eager for a final solution to the problem of "subhumans".) On the editorial board of *Der Biologe* was the Austrian ethologist Konrad Lorenz: he eagerly promoted the notion of racial purification, and recommended euthanasia for the unfit.* The quack doctor Joachim Mrugowsky (1905–1948), unable to make any impact upon academia before the rise of the Nazi regime but by 1942 appointed Commissar for Epidemics in the Ostland, had his own rationale for demanding racial purifi-

Konrad Lorenz

cation: the lesser races, according to Mrugowsky, were more vulnerable to epidemic diseases and, once infected themselves, spread the infection among the good folk of the superior races. To show this was the case, he deliberately infected large numbers of *Untermenschen* for experimental purposes, while also enthusiastically thinning their number from the population of the area under his control. He was sentenced to death at Nuremberg as a war criminal.

Meanwhile physical anthropologists like Bruno Beger and August Hirt (1898–1945) were being allowed to exploit

---

* After the war, Lorenz so efficiently remade himself that he's now primarily remembered as the animal behaviourist who received the 1973 Nobel Prize and as the author of such charming books as *King Solomon's Ring* (1949) and *Man Meets Dog* (1950). *On Aggression* (1963), risibly in view of Lorenz's past, promoted the idea that aggression, important for survival among animals, could in the human animal be constructively channelled into other things.

the concentration camps for "specimens" – skulls, bones, and so on. The concentration camp at Natzweiler, near Strasbourg, became almost a specialist site for the gassing of the "donors". These remains, along with those from captured Russian soldiers, were disseminated among various German and Austrian institutions so that researchers could explore the differences, often using such pseudo-sciences as phrenology, between Jewish or Russian anatomy and that of full humans. Of course they found the differences they were looking for.

The Romantic dream of the long-ago Aryan race lived on among many Germans after the end of WWII – not just among the latterly growing infestation of neo-Nazi skinheads the country has suffered, especially since the fall of the Berlin Wall, but also among those whose intentions are strictly pacifist. In her book *Karma Cola* (1979) Gita Mehta reports a conversation she had somewhere in India in the 1960s or 1970s – the book is consistently vague about particulars – with a Swiss who had come to India in search of enlightenment. He was disturbed by the Germans who were doing likewise. Why?

> What India does to them, what they come here to find. They are not fools like the French, they don't live in slums like the British. No. The Germans go to the mountains, the Himalayas, The Abode of Snow. And do you know what they do there? All alone, eighteen thousand feet in the air, next to a frozen glacier, with their books? They try to be Supermen.
> . . . It is an extraordinary thing about the Germans. They are like vampires. Waiting, waiting for the twilight of the gods. You know, sitting alone in the Himalayas a man can believe the gods are dying. The wind is so shrill. And the Germans wait. To take the place of the gods.

THE MEDICAL NIGHTMARE

Julius Streicher, founder and editor of the virulently anti-semitic newspaper *Der Stürmer* and later sentenced to death at Nuremberg as a war criminal, convinced himself that cancer was a bacterial infection, and demanded that German medical scientists investigate this theory. This was typical of the way that crank theories rose to the top of the broth during the dark age of the Reich. However, in thinking of the Nazi corruption of medical science, our thoughts inevitably turn first to the loathsome activities of one man: Josef Mengele, the Angel of Death.

Much of what we know about Mengele's career at Auschwitz comes from or is confirmed by the memoir *Beyond Humanity: A Forensic Doctor in Auschwitz* (1946), published shortly after the end of the war by Miklos Nyiszli (*c*1900–1956), a Hungarian Jew and physician who, as a prisoner at the death camp, had the misfortune – or perhaps fortune, because it saved him from the gas chamber – to be pressed into service during the summer of 1944 as Mengele's assistant and who consequently witnessed some of the worst savageries of this psychopath. Prisoners at Auschwitz first encountered Mengele on arrival, when he – and other physicians – selected which people were fit enough to be retained for hard labour and which should go straight to the gas chamber. Later, women might meet him again under similar circumstances, during roll calls, when some would be selected by Mengele for "special work detail". Any ideas they might have that this was a fortunate option, in that they were being spared the miseries of hard labour under the whips of the guards, were soon dispelled, because Mengele was choosing them as subjects for his experimentation, such as shock treatment and sterilization.

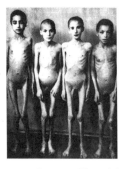

Few survived: if the experiments themselves did not kill, then infection of the untreated wounds did.

Among the more gruesome of Mengele's experiments was to inject dye into children's eyes to see if this would change their eye-colour, and the sewing of two small children back-to-back, linking their blood supplies via the veins at the wrist, in an attempt to create Siamese twins artificially. Twins were a particular preoccupation of Mengele; through experiments like this one, as well as through treating genuine biological twins differently and observing the results, he hoped to show that perceived defects and abnormalities were of racial origin – "in the blood" – rather than specific to individuals or families. In another experiment, Mengele casually murdered four pairs of twins by lethal injection purely so their eyes could be sent for study to a scientist at the Kaiser Wilhelm Institute of Anthropology who was writing a paper on the hereditary eye abnormality heterochromatism.

People suffering other congenital abnormalities, most especially dwarfs, were singled out for Mengele's special attention. Most died of his experiments or the aftermath thereof; he seems to have made a particular attempt, however, to keep alive the members of the Lilliput Troupe, seven dwarf Romanian Jews of the Ovitz family who had toured Eastern Europe as an act before being deported to Auschwitz and Mengele's tender mercies.

Mengele was far from the only physician at Auschwitz to use the opportunity of the Reich's complete abolition of normal human laws in order to experiment upon live human guinea pigs, although he seems to have worked with especial cruelty and lack of justification. "Justification" may seem an odd word to use in this context, but many of the experiments performed by the other medical scientists did at least have some purpose in terms of the war effort.

The camp was obviously an ideal place, too, to study the progress of dysentery – it was, predictably, rife among the inmates. Other diseases were brought within the physicians' ambit by infecting luckless victims appropriately and watching their decline, suffering and death. Similarly, poisons were tested. Nyiszli commented:

> Dr Wolff was searching for causes of dysentery. Actually, its causes are not difficult to determine; even the layman knows them. Dysentery is caused by applying the following formula: Take any individual – man, woman, or innocent child – snatch them away from their home, stack them with a hundred others in a sealed box car, in which a bucket of water has first been thoughtfully placed, then pack them off, after they have spent six preliminary weeks in a ghetto, to Auschwitz. There, pile them by the thousands into barracks unfit to serve as stables. For food, give them a ration of mouldy bread made from wild chestnuts, a sort of margarine of which the basic ingredient is lignite, thirty grams of sausage made from the flesh of mangy

horses, the whole not to exceed 700 calories. To wash this ration down, a half-litre of soup made from nettles and weeds, containing nothing fatty, no flour or salt. In four weeks, dysentery will invariably appear. Then, three or four weeks later, the patient will be "cured", for he will die in spite of any belated treatment he may receive from the camp doctors.

Perhaps even worse than Mengele was the SS lieutenant and physician Siegmund Rascher (d1945), a member of the Institute for Applied Research in the Natural Sciences – a grandly titled adjunct to Himmler's Ahnenerbe. For the Luftwaffe, Rascher subjected a series of live prisoners at Dachau to extreme low pressures, reproducing the effects of high altitudes. The mortality rate was about 80%. Not content with this, he and two physicians called Holzlöhner and Finke then began experimenting, again with live prisoners, into ways of resuscitating airmen who'd crash-landed at sea and spent a while in icy waters before rescue. Of course, one part of the research had to focus on exactly how long the human body could stay submerged in freezing water before death; some of the Russian prisoners lasted for hours. Out of 200 or so subjects, about 40% were tested to death. In the experiments whose purpose was revival, the subjects were dragged comatose from the water and swaddled in bedclothes, immersed in a hot bath, or – the preferred experimental technique – put to bed with one or more prostitutes, who would use their oldest-professional skills in an attempt to revive the sufferer. Sometimes they were so successful in this that the revivee was capable of sex with them, an experimental result that Rascher observed eagerly. The results were summarized by the three physicians in the paper "Concerning Experiments in Freezing the Human Organism" (1942):

> If the experimental subject was placed in the water under narcosis, one observed a certain arousing effect. The subject began to groan and made some defensive movements. In a few cases a state of excitation developed. This was especially severe in the cooling of head and neck. But never was a complete cessation of the narcosis observed. The defensive movements ceased after about five minutes. There followed a progressive rigor, which developed especially strongly in the arm muscula-

ture . . . The rigor increased with the continuation of the cooling, now and then interrupted by . . . twitchings. With still more marked sinking of the body temperature it suddenly ceased. These cases ended fatally, without any successful results from resuscitation efforts.

Holzlöhner and Finke thereafter discontinued their involvement, on the grounds that all to be learned had been learned, but Rascher carried on gleefully. He proposed moving, though, from Dachau to Auschwitz: at the larger camp there'd be less risk, he pointed out, of the screaming attracting adverse attention.

On October 25 1946 began The Medical Case, the first of the Nuremberg Trials. There were 23 defendants, of whom 20 were physicians. Of these, 16 were found guilty of experimenting on human beings without their consent, often involving severe torment, and of the cold-blooded murder of some of their victims. Seven were sentenced to death, including Mrugowsky, the Ahnenerbe chief Wolfram Sievers, and Buchenwald's chief medical officer, Waldemar Hoven. Five of those who'd played a central part in the Nazi medical experimentation were not at the trial. Rascher had been executed in 1945 on Himmler's orders, for having lied to Himmler about, of all things, his wife's fecundity. Ernst Grawitz had committed suicide before the trial. Carl Clauberg was tried in the USSR in 1948, and sentenced to 23 years; in 1955 he was released to West Germany as part of a prisoner exchange, but soon rearrested there in response to public outcry, dying of a heart attack in 1957 before his case came to trial. Horst Schumann disappeared without trace. And Josef Mengele escaped to South America, where he lived until his death in Brazil in a drowning accident in 1979.

As a gruesome footnote to a gruesome tale, we can add that many of the 200 or so physicians who'd experimented in the Nazi death camps were recruited at the end of the war, under the authority of General Dwight Eisenhower, to work for the US military, bringing the fruits of their researches with them.

Л.П.БЕРІЯ

Backdrop: Lavrenti Beria
Above: Lysenko addresses the Kremlin

STALIN'S RUSSIA

> Leave them in peace. We can always shoot them all later.
> Stalin, responding to Beria's concerns about the ideological
> soundness of the physics used in the Soviet bomb project

IT COULD ALL HAVE GONE so well. In *Science and Politics* (1973) Jean-Jacques Salomon summed up the advantages and disadvantages of Soviet-style science in a single sentence:

> There can be no doubt that scientific activities enjoyed a status and support in the USSR which had no parallel in any other country before the second world war; the scientific system was inseparable from the political system, of which it was both the means and the end.

Today, the level of governmental support for (non-military) science is more or less a measure of how developed a nation is, and it's accepted almost without question that any nation which fails to support science and science education is economically doomed, sooner or later, unless other factors are playing an extremely important part. This was the thrust of a famous US paper commissioned from electrical engineer Vannevar Bush (1890–1974) by President Franklin Roosevelt and presented in 1945 by Bush to Roosevelt's successor, Harry Truman. Even so, the only other country aside from the USSR fully to take the idea to heart before WWII was France, where in the 1930s Nobel laureates Irène Joliot-Curie (1897–1956) and then Jean Baptiste Perrin (1870–1942) were appointed to the government as Under-Secretary in order to promote and advise on science.

Wise governments support science while interfering with its conduct as little as possible – except arguably during wartime. Without US Government direction, something like the Manhattan Project could never have come into existence; but it's important to remember that the Manhattan Project is one of the few exceptions to the general rule. It's obviously legitimate for politicians to ensure taxpayer money is being spent wisely, in which exercise it's to be hoped they call upon scientific advisers to help distinguish what is genuine research from what's just pork or a wild-goose chase. However, when politicians attempt to take control of the scientific endeavour, or to tell scientists what their science *should be*, disaster is inevitable. This was emphatically exemplified in the USSR during the years when Josef Stalin, through the willing medium of Trofim D. Lysenko (1898–1976), forced the Soviet biological sciences to adopt rank pseudoscience.

HEREDITY CAST DOWN

In the USSR during the 1920s there was considerable turmoil in the field of genetics. On the one hand there were those geneticists who subscribed to ideas developed from those of the Chevalier de Lamarck (1744–1829), that evolution worked through individual living creatures adapting to environmental stimuli during their lifetimes and passing on these acquired characteristics to their offspring. On the other were those who, aware of the work done by researchers such as Gregor Mendel, realized the instrument of heredity was the gene, and that it was random mutations in the genes that drove evolution. The two schools of thought were not entirely antagonistic: compromise seemed possible, just as, in the early days after Darwin had put forward the theory that natural selection was the mechanism of evolution, there still seemed, even to him, room for Lamarckian ideas. The debate in the USSR did not differ in any important respect from that going on anywhere else in Europe, or indeed in the US.

But then a peasant amateur agronomist called Trofim D. Lysenko came upon the scene, and it was the USSR's considerable misfortune that he did so at a time when earlier foolish ideas of V.I. Lenin (1870–1924) that science should be in accordance with such Marxist and Engelian concepts as dialectical materialism were being hardened into political dogma by Stalin. A corollary of Marxism is that human nature can be moulded: such motivations as greed and ambition can be sublimated to serve society as a whole rather than just the individual. (It's the same basic tenet as that of Christianity, and indeed of the famous John F.

Kennedy speech: "Ask not what your country can do for you – ask what you can do for your country.") The essential madness of Lysenko was in thinking that the basic nature of plants (their genetic coding) could be moulded in the same way. That he should use Marxism as a guide in this and in much else was not so illogical, in a way, because under the Soviets Marxism was regarded as *itself* a science.

Lysenko's thinking did not spring from a vacuum. At an early stage in his career, posted to the remote North Caucasus, he encountered Ivan Vladimirovich Michurin (1855–1935) and was much influenced by him. Michurin, from Koslov (later named Michurinsk in his honour), was an aristocrat who tried to get along as best he could under the new, post-Revolutionary regime. In his private orchard – which was in due course turned over to the state – he worked as a fruit-grower and, his fruit-growing ambitions inhibited by the climate (he wanted to grow in the chilly Caucasus fruits that really belonged further south), in due course necessarily a plant-breeder. For philosophical reasons Michurin detested the notion of heredity, and this fitted in well with the preconceptions of the Party. Those same preconceptions turned what could have been a tragedy for this "reborn peasant" into a triumph: the Party attempted to help his efforts by sending him some genuine geneticists, who of course discovered promptly that Michurin's hybridization experiments were a shambles; he called them liars, and was lionized as the true, man-of-the-soil peasant scientist resisting elitist theoreticians and their antisocialistic falsehoods. It was a means of "debate" that Michurin's protégé, Lysenko, would use frequently during succeeding decades.

It was while working in the Caucasus that Lysenko resurrected the traditional peasant technique of vernalization. By moistening and chilling seeds in the winter one can accelerate their development after they've been planted in the spring, essentially altering the habit of winter wheat so that it behaves as spring wheat; the technique is of considerable use in high latitudes where the summers are short. Lysenko announced the technique as a great new discovery, and was chagrined when it was smartly pointed out by his peers that it was anything but. His response was that of blusterers anywhere: when a claim is challenged, rather than weigh the merits of the challenge, retaliate by making the claim ever more extravagant. In Lysenko's case, this meant insisting that vernalization could transform any strain of wheat such that it eared earlier in the year; this would obviously – obviously to Lysenko, that is – increase the yield over the year as a whole.

Despite the fact that attempts to put his principle into practice had at best mixed results, Lysenko was steadily promoted up the Party's agricultural ladder, and by 1930 he'd reached the Moscow Institute of Genetics. There he continued his vernalization experiments, extending them to cuttings, bulbs and tubers and claiming unprecedented success. It was in 1935 that he announced his new theory of heredity, which rejected Mendelian genetics entirely and instead claimed environment as the prime controller of how organisms developed. Notably, Lysenko eschewed the customary process of submitting papers on his results to scientific journals, instead preferring the medium of the newspaper interview; ever since, the use of this tactic has been taken, rightly or wrongly, as symptomatic of bad science.* This was a distinction lost on Stalin, who was desperate to hear good news – any good news, even if illusory – in the wake of a disastrous series of failed Five-Year Plans. To Stalin, Lysenko and vernalization represented the Great Hope. In 1940 Lysenko was made Director of the Moscow Institute of Genetics, a position of enormous political power that he used to destroy – sometimes literally – all those who challenged his pseudoscientific ideas; he would retain the position until his forced resignation in 1965, although mercifully his power waned after Stalin's death in 1953.

Sometime before 1936 Lysenko met a lawyer called I.I. Prezent who, despite a lack of scientific training, considered himself an expert on Darwinian evolution.† It was in conjunction with Prezent that he formulated his own theory, vernalization (in a confusing new use of the term), to replace genetics – or "Mendelism", as Lysenko preferred to call the latter. Exactly what that theory was is rather hard to

---

* Later Lysenko started and edited the journal *Yarovizatya* ("Vernalization") to use as his mouthpiece, essentially self-publishing.
† Sound familiar?

explain, because Lysenko never put forward any statement of it that actually makes self-consistent sense. In *False Prophets* (1986) Alexander Kohn offers a sample:

> The speeded up development of such plants (sprouted) we explain basically not by the fact that the eyes of the tubers are sprouted before planting, but by the fact that the sprouts are subjected to the influence of certain external conditions, namely the influence of light (a long spring day) and of a temperature of 15–20°C. Under the influence of these external conditions (and that precisely is vernalization), in the potato tuber's eyes, as they start to grow, there occur those quantitative changes which, after the tubers are planted, will lead the plant to more rapid flowering, and hence to more rapid formation of young tubers.

Kohn is being kind when he comments: "This represents an explanation without substance." It in fact is no explanation at all, just a piece of verbose woffle.

Woffle or not, the study of genetics in the USSR, although it survived in increasingly harassed form until about 1935, was thereafter ruthlessly exterminated – as were many Soviet geneticists who dared to speak out against Lysenko's increasing lock on power or the absurdity of his ideas. The vast majority of those working in the biological sciences in the USSR switched their allegiance, at least on the surface, to Lysenkoism, which may seem a cowardly action with hindsight but must have been perfectly reasonable at the time, bearing in mind the alternatives. (Besides, by this time many establishment scientists were not recognizably scientists at all, being political appointees chosen for their loyalty to the Cause rather than any competence.) Others were made of sterner stuff. Those holdouts who were lucky were forced from their jobs. Many more were arrested and either were executed or died in prison or concentration camp.* Among the distinguished geneticists persecuted were Nikolai Ivanovich Vavilov (1887–1943), G.D. Karpechenko (1899–1941), Salomon Levit, Max Levin and Israel Agol; even the US geneticist H.J. Muller (1890–1967), a committed communist who'd taken up a

---

* It's hard to establish a figure, but probably approaching 100 were "disappeared" in this way.

post in 1933 at the Institute of Medical Genetics,
Leningrad, and later (1946) to be a Nobel laureate, felt
sufficiently threatened that he fled the country. Of course,
these represent just a tiny fraction of the deaths that can be
laid at Lysenko's door: it's estimated that something like 10
million people died as a consequence of the lethal combination of the collectivization of Soviet agriculture and the
imposition of Lysenkoist agricultural practices, either
through starvation in various famines or through violence
in Stalin's extermination campaigns against those farmers
who pointed out that the "improvements" weren't improvements at all.

What is truly appalling is that all of this evidence of the
complete uselessness of Lysenkoist ideas and techniques
must have been perfectly obvious to Stalin and the Party
bureaucracy, and yet, so blinded were they by the fact that
Lysenkoism seemed ideologically sound, they not only
permitted him to continue but increased their support for
him. If reality stubbornly refused to conform to Marxist
science, then it must be reality that was at fault.

Lysenko did not restrict his pseudoscience to such practical matters as agriculture and breeding, where his destruc-
tive influence and his wild theorizing were far more widespread than can be indicated in the brief sketch above. He
also, from 1948, began to mount a concerted attack on the
theory of evolution by natural selection, perhaps in hopes of
toppling Darwin from his pedestal and installing himself
there instead. In the natural world, declared Lysenko, there
was no such thing as competition for survival between individuals within a species, since such a concept violated
Marxist principles of cooperation. Thus the basic idea that
new species arise from old ones through chance advantageous mutations within the species giving some individuals
advantage over others was patently ridiculous. New forms
arose through crossing between one species and a different
one . . . and he and his colleagues had done the experiments to prove it, transforming wheat into rye, cabbage into
swede, barley into oats, and so on. None of these experiments were ever properly written up for publication, and
none of them proved replicable by other scientists.

Another oddity of the Lysenkoist scheme was that he
refused point-blank to believe in the existence of hormones.

Oddest of all was his insistence that cuckoos were not really a distinct species, laying eggs in the nests of song-birds of other species; rather, cuckoo chicks emerged from eggs laid by song-birds who'd taken to eating caterpillars.

After Stalin's death in 1953, Lysenko continued to enjoy the support of Stalin's successor, Nikita Khruschev (1894–1971), although there was the sense that this support was steadily ebbing. Throughout the 1950s, agricultural science was progressing by leaps and bounds in other countries of the world, and it was impossible for the Party bureaucracy to ignore the fact that many of these developments were happening precisely because the foreign scientists were working directly in contradiction to the version of reality preached by Lysenko. As an example, Lysenko and his adherents declared it impossible to stimulate superovulation in sheep through the use of hormones, yet the Soviet Minister of Agriculture witnessed exactly this being done during a visit to the UK, and, despite Lysenko's protestations, the technique was thereafter introduced (in fact, reintroduced) to the USSR. Even earlier, in 1955 the Soviet Academy of Sciences became sufficiently confident to elect as head of its biology division a real biochemist, V.A. Engelhardt (1894–1984), and he soon set to work bringing genetics back into the fold of Soviet science.

After Khruschev was deposed in 1964, Lysenko's fall from grace accelerated. In 1965 the Ministry of Agriculture and the Soviet Academy of Sciences convened a panel to examine a particular claim of Lysenko's in relation to the breeding of cattle with a high buttermilk yield, and the panel soon concluded that Lysenko and his confreres had been conducting scientific fraud on an almost unimaginably grand scale. Lysenko was forced from his position as Director of the Institute of Genetics of the Soviet Academy of Sciences, a position he'd held since 1940, and of the Lenin Academy of Agricultural Sciences. Over the next couple of years Lysenkoist ideas were removed from science education, being replaced where appropriate by Mendelian and Darwinian ones. It was perhaps a further quarter-century before Russian genetics caught up with that in the rest of the world, and indeed it can be argued that this has yet to happen.

Lysenko was not the only Russian geneticist to have his

dubious scientific hypotheses ludicrously over-promoted under Stalinism. Before him there had been the sad case of Ilya Ivanovich Ivanov (1870–1932). Ivanov earned his reputation under the Tsar as an animal breeder, specifically for his work on the artificial insemination of racehorses. He also studied the use of artificial insemination to produce inter-species hybrids, and succeeded in creating a zeedonk (zebra/donkey), a zubron (European bison/domestic cow) and various other crosses. Even before the Russian Revolution, he was interested in the possibilities of using similar techniques to produce ape/human hybrids; in 1910 he gave a presentation in Graz to the World Congress of Zoologists describing exactly such a scheme.

After the beginning of Stalin's reign in 1924, new opportunities to pursue this research opened up for Ivanov. Exactly who the notion originated with is unclear, but Stalin was entranced by the idea that hybridization between apes and humans held the potential of breeding for the Red Army a new race who would be both uncomplaining workers and super-strong warriors, unafraid of death and untempered by compassion – a "living war machine", as the Politburo described the ideal. After some false starts, in 1927 Ivanov began trying to impregnate female chimpanzees with human sperm (from donors; not his own!) at the zoological gardens in Conakry, French Guinea.

He brought back from Conakry to the USSR – in fact, to his new primate centre in Sukhumi, Georgia – a collection of apes including two of the three chimps he had attempted to impregnate (the third died *en route*). Meeting no success at Sukhumi, as in Conakry, with the impregnation of chimp females, his next approach was to attempt to inseminate human females with ape sperm. In 1929 he gained the support of the Society of Materialist Biologists for such an experiment, and five volunteer women were lined up. Before he could begin, however, his only surviving sexually mature ape at Sukhumi died, and he had to start the slow process of importing some replacements. These arrived in the summer of 1930, but by then Stalin was imposing a general crackdown on Soviet biologists in consequence of the repeated failure of various agricultural plans, and Ivanov was not immune. Arrested late in 1930 he was sentenced to five years' exile in Kazakhstan. There he worked for the Kazakh Veterinary–Zoologist Institute until dying of a stroke in March 1932.

Ivanov was deluded but harmless. Not so Olga Borisovna Lepeshinskaya (1871–1963),* who was very much in the Lysenko mould. An unqualified researcher, she was promoted because of her peasant credentials and because the type of "science" she expounded was politically accept-able in the Stalinist context; the fact that she had been a friend of Lenin's didn't hurt. Her earliest (vaguely) relevant experience was to study midwifery in St Peters-burg, finishing her course in 1887. Not much is known about her until 1915, when she completed a medical course for women and was appointed an assistant lecturer in the University of Moscow's

Lepeshinskaya in later life

Department of Medicine, where she stayed, with a brief sojourn in 1919 to the University of Tashkent, until 1926. Next she joined the K.A. Timiryazev Institute of Biology as an histologist. In 1941 she became head of the Department of Live Matter of the Institute of Experimental Biology of the Academy of Medical Sciences, a post she held almost until her death. It was a position of great power, which like Lysenko she used ruthlessly against her scientific adversaries.

Lepeshinskaya had an early position under Alexander Gurwitch (see page 95), whose initial fall from Party favour she engineered in the late 1920s. Lepeshinskaya persuaded

---

* Not to be confused with the ballerina Olga Lepeshinskaya (b1916), a star of the Bolshoi.

herself that a phenomenon she'd "discovered" offered an alternative explanation for Gurwitch's mitogenetic rays, and so a clash between the two was inevitable.

It was well known that strong ultraviolet radiation killed living cells. Quasiparadoxically, though – at least according to Lepeshinskaya's claims – weak doses of ultraviolet instead stimulated the cells. Reasoning further, she hypothesized that, when cells died, they must re-release this energy in the form of ultraviolet light; and, *mirabile dictu*, this was precisely what her experiments showed. She investigated this wonderful new form of radiation for some while, although most Soviet biologists of the time seem to have given the whole subject a wide berth. A few years later, once she had risen to power, such disregard was a luxury few of them were permitted. By now she had made a further remarkable breakthrough in the field of microbiology: it was known that cells reproduced by binary fission (splitting into two identical halves), but Lepeshinskaya was lucky enough to observe another means of cell propagation. In this mode, instead of dividing, they disintegrated into granules, and these granules could then reconstitute themselves into cells – all sorts of different cells, not just the original cell-type. Even better, crystals of inorganic substances could, by the application of nucleic acids, be turned into living cells. The fundamental unit of life, then, she reasoned, was not the cell but a sort of disorganized "living matter" that could be manipulated however the skilled researcher (herself) wanted to manipulate it.

It was only a short step from this latter "discovery" to the notion of spontaneous generation.* She ground up microbes and left the dead residue in a flask for a few weeks. By then there were plentiful signs of life in the flask thanks, although Lepeshinskaya did not know this (presumably having consigned Pasteur, along with so many other figures of scientific history, to the trashcan of history), to airborne microbial spores. Soon the Lepeshinskaya-inspired renaissance of spontaneous-generation ideas permeated much of Soviet biological sciences with, as might be expected, highly

---

* The long-held idea, debunked decades earlier by Louis Pasteur, that living organisms could spring into being from inorganic materials or just out of nowhere.

placed sycophantic quasi-scientists reporting experiments that not just confirmed her views but also extended them into new areas of lunacy – for example, that you could grind up pearl buttons and grow living tissue from them by injecting them into animals.

In 1950 the Biological Sciences section of the Academy of Sciences held a symposium called "Live Matter and Cell Development". The chairman was the genuinely distinguished biochemist A.I. Oparin (1894–1980),* who must certainly have known that what he was listening to was purest tommyrot, but who throughout the Stalinist years seems to have been more concerned with preserving his own skin than anything else. Lepeshinskaya gave the keynote speech, a political tirade much approved by Lysenko. The symposium declared Lepeshinskaya's researches to be indeed of revolutionary importance; she was elected a Member of the Academy of Medical Sciences and awarded the Stalin Prize. Teachers in medical faculties throughout Russia were instructed, under fear of penalty, to refer to her doctrine in every lecture; any favourable reference to the reality accepted in the rest of the world, that cells arose from other cells via division, was outlawed.

Loren Eiseley (1907–1977) summed the situation up in *The Immense Journey* (1957):

> Every so often one encounters articles in leading magazines with titles such as "The Spark of Life," "The Secret of Life," "New Hormone Key to Life," or other similar optimistic proclamations. Only yesterday, for example, I discovered in the *New York Times* a headline announcing: "Scientist Predicts Creation of Life in Laboratory." The Moscow-datelined dispatch announced that Academician Olga Lepeshinskaya had predicted that "in the not too distant future, Soviet scientists would create 'life.'" "The time is not far off," warns the formidable Madame Olga, "when we shall be able to obtain the vital substance artificially." She said it with such vigor that I had about the same reaction as I do to announcements about

---

* Oparin's great contribution was his hypothesis concerning the origin of life, put forward in 1924 and summarized in *The Origin of Life on Earth* (1936) and others. His ideas were picked up and developed in the 1950s by the US chemist Harold Urey (1893–1981), and held considerable sway for some decades.

atomic bombs. In fact I half started up to latch the door before an invading tide of Russian protoplasm flowed in upon me.

What finally enabled me to regain my shaken confidence was the recollection that these pronouncements have been going on for well over a century. Just now the Russian scientists show a particular tendency to issue such blasts – committed politically, as they are, to an uncompromising materialism and the boastfulness of very young science.

After Stalin's death, reality was allowed to creep back into Soviet cytology. Lepeshinskaya and her ideas were never dramatically denounced, as happened with Lysenko; rather, they were allowed to slide into obscurity, as was she.

DIALECTICAL MATERIALISM DEFINES PHYSICS

Two things saved Soviet physics under Stalin from the same sort of devastation that Soviet genetics suffered during that era: Stalin's enormous respect for the physicist Pyotr Kapitsa (1894–1984), and the Soviet atom-bomb project.

Almost from the start, Soviet physics was in a quandary over the new reality portrayed by Einstein's Relativity, which seemed to fly in the face of dialectical materialism – i.e., "common sense".* Dialectical materialism prescribed, to simplify, that physics should be, well, physical: the universe, its structure and its laws should all be solidly within the realm of the human senses. A concept like the luminiferous aether, the supposed substrate of the universe, sat comfortably within the limits of dialectical materialism because, even though the aether was proving infernally hard to detect, it was at least theoretically detectable: one could envisage that, somewhere down the line, it would be brought within the grasp of the tangible. Yet a basic tenet of the new physics represented by Relativity was that the aether didn't exist. Just to confuse matters for the ideologically conscientious Soviets, no less an authority than Engels

---

* In general, Soviet physicists quietly ignored all the public political hoohah and with minimal interference carried on their work, which was fully acceptant of Relativity and, soon after, quantum mechanics; throughout the Stalinist era, Soviet physics ranked among the world's best. Physicists in Nazi Germany were not nearly so lucky.

thoroughly approved of Relativity, which he felt was in full accordance with dialectical materialism since it gave geometry a materialist expression.

Fortunately, there was a way out of the dilemma. Lenin had condemned the notion of Ernst Mach (1838–1916) that our understanding is limited by our perceptions – by what we are able to observe – so that the fundamental *causes* of what we observe will always remain a mystery to us. Relativity could be portrayed as a description of *how* things behaved rather than an explanation of *why* they behaved that way, and therefore could be denounced in staunch Leninist terms as "Machist" – which a number of Soviet physicists proceeded to do, some of them also echoing the antisemitic denigrations of Relativity by the Nazi physicist Philipp Lenard (see pages 245ff). These denunciations were countered by other physicists who claimed, like Engels, that Relativity was in full accordance with dialectical materialism. That this argument in these terms should have been taking place at all is symptomatic of the corruption of science in totalitarian regimes, a corruption brought about by the irrational compulsion to distort reality until it can be made to fit a preconceived mould.

Of course, matters got much stormier when quantum mechanics exploded on the scene. A concept like Heisenberg's Uncertainty Principle could be seen as a calculated slap in the face for dialectical materialism. Even worse was when Lysenko decided in 1948 that, having conquered Soviet genetics, he should make his mark on Soviet physics as well. His influence, although the intellectual basis of his dogma was puerile, probably did more damage than all that had gone before. According to Lysenko, Soviet physics represented the true path while all foreign physics, with its funny ideas about Relativity and quanta, was an insidious threat to the Glorious Revolution, a deliberate attempt by the corrupt, decadent West to undermine the USSR. Soviet physicists were therefore forbidden to maintain their contacts with their counterparts in the West, and accordingly – especially since those who disobeyed Lysenko had an unfortunate habit of disappearing – Soviet physics withdrew into its shell. It was fortunate for the nation that Soviet physics was at the time sufficiently strong that, even despite the isolation, it could remain among the world-leaders.

In a celebrated instance, the *Soviet Encyclopedia* published an article that was stoutly anti-Relativity, proclaiming that the aether existed and that Newtonian physics was perfectly adequate to explain the universe. The physicists George Gamow (1904–1968), Lev Landau (1908–1968) and M.P. Bronstein (1906–1938) wrote facetiously to the encyclopedia's editor saying they accepted the dictum about the aether's existence and sought similar guidance concerning such substances as caloric and phlogiston.* Bronstein and Landau were promptly kicked out of Moscow University, while Gamow, already too notable for such treatment, suffered later when, while lecturing, he mentioned Heisenberg's Uncertainty Principle; a local political officer stopped the lecture midflow and dismissed the audience, Gamow himself receiving a disciplinary warning. In 1934 Gamow left for the US, where he became Professor of Physics at George Washington University; he spent the rest of his life working in the US. Landau and Bronstein were not so lucky, the outspoken Landau in particular being periodically hounded by the authorities until Stalin's death in 1953; more than once Pyotr Kapitsa, who had Stalin's ear, had to come to Landau's rescue.

Kapitsa, born in Kronstadt, studied first in Petrograd and then in England, at Cambridge under Ernest Rutherford (1871–1937); he served as Assistant Director of Magnetic Research at the Cavendish Laboratory, Cambridge, from 1924 to 1932. In 1934, while on holiday in his native Russia, he was abruptly seized by the authorities under Stalin's orders, and installed as Director of the Institute of Physical Problems, Moscow. There was of course much international outrage, to which Stalin was deaf; notable among those who declined to protest was, oddly, Rutherford, who declared the USSR needed Kapitsa more than the West did and arranged for some of Kapitsa's equipment to be shipped to him. Stalin seems to have revered Kapitsa as a kind of scientific god, because time and again Kapitsa was able to intercede on behalf of one scientist or another who was suffering threats for straying publicly from the path prescribed by dialectical materialism.

---

* Both long before consigned to science's dustbin of discarded ideas.

In return, Kapitsa was a good servant of the USSR, not only as the living flagship of Soviet physics but also as a valuable technological innovator. It was his favoured status with Stalin that almost certainly saved his life when, in the wake of the US destruction of Hiroshima and Nagasaki, the project was initiated to produce a Soviet atomic bomb. Kapitsa seems to have been reluctant from the outset, but he became

Kapitsa

even more so when the man put in overall charge of the project proved to be none other than Lavrenti Beria (1899–1953), the dreaded boss of the KGB. The choice could hardly have been more idiotic, and Kapitsa told Stalin so, requesting that Beria be replaced. Maddened by the insubordination, Beria demanded Kapitsa should suffer the same fate as others who'd defied him. Instead, Stalin put Kapitsa under house arrest, where he remained until Stalin's death and Beria's execution in 1953.

The bomb project itself could have been nullified by the Lysenko infection. One of those particularly influenced by Lysenko's ideas concerning bourgeois Western physics was Sergei Kaftanov, Stalin's Minister for Advanced Education. In March 1949 he planned to summon a committee with the intention of outlawing Relativity and quantum mechanics from Soviet physics forever. For once Beria had the good sense to take a second opinion on this, from the project's director and head of the Soviet Atomic Energy Institute, Igor Kurchatov* (1903–1960), who told him bluntly that, without Relativity and quantum mechanics, Beria's pet baby wouldn't explode. After consultation with Stalin (who responded with the remark at the head of this subchapter),

* Also, much later, in 1954, a leading figure in the creation of the world's first industrial nuclear power plant.

Beria cancelled Kaftanov's committee and the atomic scientists were allowed to proceed. Just a few months later, in August 1949, the USSR successfully detonated its first atomic bomb.

THE LUNATICS CONQUER THE ASYLUM

In Russia the use of false psychiatric diagnoses of political dissidents, and their consequent incarceration in mental institutions as a means of silencing them (out of sight, out of mind, as it were) that avoided the potential political embarrassment of imprisoning, executing or "disappearing" them, dated from long before the Stalin regime and indeed the Revolution, and probably hit its heights after Stalin's death, during the reign of Nikita Khruschev; this may not reflect so badly on Khruschev as at first appears, but simply be a manifestation of the fact that Stalin was less concerned about "political embarrassment" than his successors. It is likely this false hospitalization continues even today.

The earliest known Russian dissident to suffer thus was Pyotr Chaadayev (1793–1856), classified as deranged by Nicholas I (1796–1855), who thereby simultaneously nullified Chaadayev's previously expressed dissidence and had the man stilled from expressing any more of it. The practice seems to have been repeated on occasion both before and after the Revolution: initially the Bolsheviks seem to have been no worse offenders in this respect than the Tsars before them.

It was certainly used against Maria Spiridonova (1884–1941), leader of the rival Left-Socialist Revolutionary Party and thus seen as a threat; she had been savagely tortured and imprisoned in Siberia in 1905, for the assassination of a brutal police chief, during the earlier, unsuccessful Russian revolution, then released in 1917. After her release she became more popular than ever. Following the collapse of the alliance between the Bolsheviks and the Left-Socialists in 1918, she was imprisoned and tortured twice more in rapid succession – *plus ça change* – all of which merely increased the number and fervour of her followers. A Moscow Revolutionary Tribunal of 1919 saw a way out of the impasse, declaring that she should be put in a sanatorium for a year to undergo mental and physical therapy. In the event, she avoided this.* In 1922 a similar technique was tried on the dissident Angelica Balabanov (1878–1965), who had served briefly as Secretary of the Comintern. A Ukrainian by birth and a resident of Italy before coming to Russia at the time of the Revolution, she was able to escape back to Italy, where in due course she helped organize resistance to Benito Mussolini (1883–1945).

These were isolated incidents, however. After the ascent of Stalin the practice became more

Maria Spiridonova

widespread, with one psychiatric hospital at Kazan being run by the NKVD and seemingly being devoted entirely to patients – often genuinely mentally ill – whose mental illness "took a political turn". Mentally sound people incarcerated at Kazan purely on grounds of political dissidence had often arrived there not through the repression of the State but through the subversive mercy of psychiatrists, who knew what the alternative to a diagnosis of derangement was likely to be. Occasionally even the police quietly colluded in this. The poet Naum Korzhavin (b1925), although eventually sent to Siberia, described how the assessing psychiatrists at the Serbsky Institute for Forensic Psychiatry did their utmost to find any reason for declaring him mentally ill; he should have cooperated with them, as he later discovered.

Not all incarcerations were so benignly intended, though: far from it. And the practice seems to have acceler-

* But hers was no happy-ending story. In 1937 she was sentenced to 25 years in prison. In 1941 she was one of a group of some 150 prisoners massacred by the NKVD in the Medvedevsky Forest.

ated toward the end of Stalin's long tyranny. One victim of it was the loyal but honest communist Sergei Pisarev, who spent a while first at the Serbsky Institute and then in the Leningrad Prison Psychiatric Hospital during 1953–5. On his release, emboldened by the fact that Stalin was gone and the less brutal Khruschev installed in his place, Pisarev went public about the scandal of habitual misdiagnosis as a means of suppressing dissenters. The Central Committee ordered a commission of inquiry, but nothing was ever done in response to the commission's report. Among subsequent victims were the writer Valeriy Tarsis, whose novel *Ward 7* (trans 1965) fictionalized his experiences, and the mathematician Alexandr Volpin (b1924), who was forcibly hospitalized no fewer than five times between 1949 and 1968. Pisarev tried again in 1970 to get something done about his nation's shame, this time taking his protest to the Academy of Medical Sciences; again nothing was done.

In 1965 the West became aware of the Soviet abuse of psychiatry, first through the UK publication of Tarsis's novel and then through the case of the interpreter Evgeny Belov, which was taken up by Amnesty International.* In the USSR itself in 1968, amid a more general rising of sentiment concerning the regime's human-rights abuses, the abuse of psychiatry was particularized, and the Action Group for the Defence of Human Rights was established. Even this dragging of the malpractice into the glare of publicity seems to have done nothing to decelerate the abuse: of the Action Group's 15 founder members, four were in due course themselves forcibly hospitalized.

Fortunate dissidents were sent to ordinary psychiatric hospitals, run by the Ministry of Health, where their care and access by friends and family were much as for genuine mental patients. By contrast the Ministry of Internal Affairs ran the dreaded "special" psychiatric hospitals, which were in many ways indistinguishable from torture camps; those

---

* Belov's case came to light purely by chance. In 1964 a group of UK students had visited the USSR and become friendly with their interpreter. They made another visit in 1965, and not unnaturally asked after their old pal Evgeny. On reaching home they told the truth to the world, campaigning for his release.

who disappeared into these were effectively cut off entirely from the outside world. Their companions in incarceration included many of the USSR's most violently and dangerously insane: rapists, murderers, arsonists. The orderlies were criminals working to fulfil their prison sentences: no one would reprove them if they behaved with extreme brutality toward those under their care, and they frequently did. The hospital directorship was a civil appointment, usually of a reliable Party hack. The medical staff might be genuine enough, but they had little chance of ameliorating conditions for their "patients" because of the prison-like structure of the establishment; they had no means of control over the other personnel, no way of stopping the torments. Copious drugs were administered primarily for the unpleasant side-effects – that is, as just another tool of torture. Particularly notorious was the drug sulphazin, which had been abandoned as therapy in the 1930s when found to have no therapeutic value that could possibly counter the intense and enduring pain of its administration.

The only escape from these hells, assuming one survived them – hopefully with mind intact – was to recant, firstly by "admitting" one was genuinely ill and then by denouncing one's earlier dissidence and claiming it to have been the product of a deranged mind.

While indisputably there were worse medical horrors perpetrated in the Nazi concentration camps, the nightmare of the Soviet abuse of psychiatry was that it perverted the medical science itself to make it a weapon of repression and torture.

Top: Cover of Michael Crichton's *State of Fear*, a bestselling
denial of anthropogenic climate change
Backdrop: Part of the icecap Crichton claims we're not melting
Above: Optimistic 1947 ad for DDT

BUSH'S AMERICA

————————⟨⟨⟨⟨⟩⟩⟩⟩————————

That's not the way the world really works anymore. We're an
empire now, and when we act, we create our own reality.
 Unnamed Senior Bush Administration Official,
 cited by Ron Suskind in "Without a Doubt",
 New York Times, October 17 2004

In short, it is prudent to regard the committed and politically
ambitious parts of the anti-science phenomenon as a reminder
of the Beast that slumbers below. When it awakens, as it has
again and again over the past few centuries, and as it undoubtedly will again some day, it will make its true power known.
 Gerald Holton, *Science and Anti-Science* (1993)

H OLTON'S WORDS ABOVE have an eerie ring for US readers
 today, after several years in which Holton's "anti-
science phenomenon" has been embedded deep within the
highest offices of government, scientific findings that
conflict with the ideological or commercial preconceptions
of the Administration being either rejected or suppressed –
or even falsified. What follows can be at best a sketch; there
has, quite simply, been far too much corruption of science
by this Administration for a single chapter to cover it all, or
even touch upon it all. Hopefully the examples given here
will serve to indicate the general, devastating pattern –
snapshots, as it were, of an immense catastrophe. Readers
seeking a fuller account are referred to Chris Mooney's *The
Republican War on Science* (2005) and Seth Shulman's
Undermining Science (2006).

THE INCONVENIENT TRUTH

Of all the areas in which the Bush Administration has
abused science, the one that has received the most attention
is global warming. For a number of years it has been the
consensus of the international climatological community

that the global climate is approaching a catastrophic change because of the release of greenhouse gases through the burning of fossil fuels. There may be other contributors towards this approaching catastrophe, but the human contribution is the only one we can do anything about. There is no question that this is the case. There are no dissenters among qualified climatologists except a few mavericks – "few" being a misleadingly strong word for a number that is so small as to be statistically invisible.

This fact is inconvenient for those industries whose livelihood currently depends on our burning fossil fuels, such as the oil and automobile corporations. It is therefore inconvenient also for politicians whose political or personal welfare depends on the financial contributions of those corporations. Ironically, many corporations outside the US, and increasingly within the country, have looked at the need for humankind to find alternative sources of energy and realized there are healthy profits to be made. Yet dinosaur corporations worldwide have entered a state of denial about the harshness of the situation, and, like spoilt children told to stop playing with a live grenade, have mounted vociferous protests against reality – as if reality, like politicians, could be swayed by bullying or bribery.

If there's essentially no question at all among climatologists as to this reality, inevitably, as with anything complex, there's debate over the details. Is the tipping point where catastrophe becomes inevitable years in the future or decades? How great is the current acceleration of glacial melting? How high will the waters rise? What will be the rate of desertification? Is the death of the Gulf Stream a matter of certainty or is a reprieve feasible? All of these discussions concern not whether the outlook is grim or optimistic but just the degree of grimness.

Nevertheless, the Bush Administration has distorted the picture of the situation it presents to the US public by pretending these debates over details are debates over substance. "The jury is still out on climate change" has been a constantly recurring, and completely dishonest, catchphrase. As a consequence the majority of the US public, unlike their counterparts throughout the rest of the world, were until very recently unaware of just how bleak the future is likely to be unless effective – and drastic – action is taken almost immediately. There has thus been little public spur to force the legislators of Senate and House of Representatives to curb those of their actions that, crazily, actually promote increases in the consumption of fossil fuels, let alone to do what is urgently required, to introduce legislation that will reduce the consumption through such measures as mandatorily enforced fuel economies in new automobiles and, most necessarily of all, initiating a quest to find new, renewable, non-polluting energy sources.

It would be bad enough were the members of the Bush Administration to confine their corruption of climate science to campaign speeches and the like. It *is* bad enough that through their legislative actions they make the situation more parlous, not less. But they go further, using Stalinist techniques to distort or suppress the climate science they find distasteful in the reports produced by their own scientific advisory bodies.

One of the worst such incidents occurred in 2003 when the White House intervened to make a number of fundamental changes to a *Draft Report on the Environment* that had been produced by the Environmental Protection Agency (EPA). Some of these were:

• Removal of all references to the 2001 report *Climate Change Science: An Analysis of Some Key Questions*, produced for the Administration by the National Academy of Sciences Commission on Geosciences, Environment and Resources. This report, which in strong terms supported the findings of the Intergovernmental Panel on Climate Change (IPCC) that climate change caused by human activities is imminent and its effects on human civilization will be catastrophic unless measures are taken immediately to ameliorate them, was effectively buried on its release through Administration members grossly misrepresenting its conclusions. Now they insisted that mention of it be expunged from the EPA report.
• The removal of a record covering global temperatures over the past 1000 years. Instead, according to a despairing internal EPA memo dated April 29 2003 on the changes demanded by the Administration: "Emphasis is given to a recent, limited analysis [that] supports the Administration's favored message."
• Still quoting from that memo, "The summary sentence has been [deleted]: 'Climate change has global consequences for human health and the environment.' . . . The sections addressing impacts on human health and ecological effects are

deleted. So are two references to effects on human health. . . . Sentences have been deleted that called for further research on effects to support future indicators."
• The Administration inserted reference to the 2003 paper "Proxy Climatic and Environmental Change of the Past 1000 Years" which, although it had appeared in the journal *Climate Research*, was in large part funded by the American Petroleum Institute. This paper's conclusions had been comprehensively discredited in the literature and thus, quite correctly, ignored by the EPA.
• Again from the internal EPA memo: "Uncertainty is inserted (with 'potentially' or 'may') where there is essentially none. For example, the introductory paragraph on climate change . . . says that changes in the radiative balance of the atmosphere 'may' affect weather and climate. EPA had provided numerous scientific citations, and even Congressional testimony by Patrick J. Michaels, to show that this relationship is not disputed."
• "Repeated references now may leave an impression that cooling is as much an issue as warming."

The EPA discussed internally the best way of coping with these distortions of scientific fact, and eventually concluded that all it could do was delete entirely its report's section on climate change: anything else would grossly misrepresent the truth and make the EPA and its scientists subject to ridicule. The omission had what one assumes must have been the desired effect: it drew attention to the fact that the White House had, in effect, attempted to make the EPA a mouthpiece for its own ideology. Not just scientists but politicians – Republicans and Democrats alike, although the Administration was swift to smear them as "partisan" – criticized the Administration's actions; among the most prominent Republican critics was Russell Train, who had been the Administrator of the EPA under two Republican presidents, Nixon and Ford. The Republican EPA Administrator whom the Bush White House had itself appointed, Christine Todd Whitman (b1946), soon departed in apparent disgust at the whoring of the truth she had been expected to institute.

The emasculation of the EPA report is, unfortunately, just one example among many. In March 2005 Rick Piltz resigned from the Climate Change Science Program, accusing that political appointees within that body were acting so as to "impede forthright communication of the state of climate science [and attempting to] undermine the credibility and integrity of the program". On June 8 2005 the *New York Times* published documents from Piltz's office which had been edited by Philip A. Cooney, a lawyer without scientific training who had represented the oil industry in its fight to prevent restrictions being placed on the emission of greenhouse gases and who was by now, incredibly, Chief-of-Staff of the White House Council in Environmental Quality. The effect of Cooney's edits was, as the *New York Times* summarized, "to produce an air of doubt about findings that most climate experts say are robust". In one instance he had removed from a report on the projected impact of global warming on water supplies and flooding a paragraph outlining likely reductions of mountain glaciers; his marginal note stated that the paragraph was "straying from research strategy into speculative findings/musings".

That same day, June 8, Bush spokesman Scott McLellan denied charges that the Administration was trying to make climate change, and the human contribution to it, appear a matter of scientific uncertainty rather than known reality. It would seem that McLellan must have been given false information. He also claimed that it was standard procedure for political appointees to edit government-sponsored scientific reports before their publication, which was surprising news to many. Two days later Cooney resigned as Chief-of-Staff of the White House Council in Environmental Quality, but he didn't stay unemployed for long: within days he was hired by Exxon Mobil.

A new outcry arose in the early days of 2006 when the distinguished NASA scientist James E. Hansen (b1941) went public with complaints that his discussions of climate science were being impeded and censored by political appointees within the NASA administrative staff. This was not the first time Hansen had been at the centre of a storm

Hansen

over the politically motivated censorship of science: back in 1988, under the George H.W. Bush Administration, he had testified before Congress that he was 99% certain that long-term global warming, probably owing to a greenhouse effect caused by overuse of fossil fuels, had already begun, and that the time to start doing something about it was now. In 1989 he discovered that this testimony had been "edited", presumably at the behest of the oil and automobile industries, by the Office of Management and Budget to stress "scientific uncertainties". Now he was up against a perhaps even worse case of censorship, this time motivated not just by political but also, incredibly, by Christian fundamentalist considerations. He had been told by NASA officials that the Public Affairs staff had been instructed to review his forthcoming lectures, papers, newspaper requests for interviews, and the like. This was because of Hansen's insistence on speaking about the urgent subject of climate change – a not unreasonable insistence, since his NASA job was to direct the computer simulation of global climate at the Goddard Institute.

Initially the NASA administrative staff came out fighting. Dean Acosta, Deputy Assistant Administrator for Public Affairs, said there were no special restrictions on Hansen that did not apply to all NASA scientists: scientists were free to discuss scientific findings, but policy statements should be left to policy makers and to appointed political spokesmen – a position that might seem understandable until you start wondering who decides where the line is drawn between scientific and policy statements. Clearly Hansen's justified warnings about climate change were being classified as the latter. According to the *New York Times*'s Andrew C. Revkin, reporting on the issue on January 29 2006 as part of his excellent series of articles following the whole fracas:

> The fresh efforts to quiet him, Dr Hansen said, began in a series of calls after a lecture he gave on Dec. 6 at the annual meeting of the American Geophysical Union in San Francisco. In the talk, he said that significant emission cuts could be achieved with existing technologies, particularly in the case of motor vehicles, and that without leadership by the United States, climate change would eventually leave the earth "a different planet."

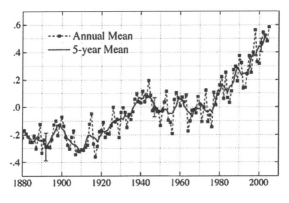

Global surface temperatures since 1880

Very soon attention focused on George C. Deutsch, a 24-year-old public affairs officer at NASA HQ, a Bush appointee whose sole qualifications appeared to be that he had been a part of the 2004 Bush–Cheney re-election campaign. In one specific instance, when National Public Radio wanted to interview Hansen, Deutsch refused permission: his reasoning was, he told a Goddard Institute staffer, that NPR was the country's "most liberal" media voice, while his, Deutsch's, job was to "make the President look good". A few days later it was revealed that Deutsch had instructed the designer of the NASA website to add the word "theory" after every mention of the Big Bang:

> It is not NASA's place, nor should it be, to make a declaration such as this about the existence of the universe that discounts intelligent design by a creator. . . . This is more than a science issue, it is a religious issue. And I would hate to think that young people would only be getting one-half of this debate from NASA. That would mean that we had failed to properly educate the very people who rely on us for factual information the most.

This is NASA, remember?

Revkin went on to report that scientists at the climate laboratories of the National Oceanic and Atmospheric Administration, accustomed in earlier years to taking journalists' phone calls whenever they wanted, were now permitted to do so only if an interview had been given the green light by Washington and only if a public affairs officer was able to monitor the conversation. Meanwhile, those rare government science employees who disagreed about the need to curb fossil-fuel emissions were permitted to lecture and publish at will.

To his great credit, the Chairman of the House Science Committee, Sherwood Boehlert, despite being a Republican and therefore expected to toe the political line, backed Hansen: "Political figures ought to be reviewing their public statements to make sure they are consistent with the best available science. Scientists should not be reviewing their statements to make sure they are consistent with the current political orthodoxy." (Sadly, Boehlert retired from Congress in 2006.) Unsurprisingly, not all of his Republican colleagues concurred; a spokesman for Oklahoma Senator James Inhofe (b1934) stated: "It seems that Dr Hansen, once again, is using his government position to promote his own views and political agenda, which is a clear violation of governmental procedure in any Administration." Read that again: Hansen's "*political* agenda"?

This was to become a theme in the Administration's attempts to smear Hansen rather than correct what was obviously an iniquitous situation. Not long afterwards, it was discovered by Nick Anthis of the Scientific Activist blog (www.scientificactivist.blogspot.com) that George Deutsch, the youthful appointee whose actions had been Hansen's final straw, had "exaggerated" his curriculum vitae a little when applying for the NASA post. (Another recurring theme among Bush political appointees; Michael Brown had similarly "exaggerated" his career when joining FEMA.) Within hours Deutsch resigned his position. Speaking the day afterwards, February 9, on Radio WTAW-AM, he said:

> Dr James Hansen has for a long time been a proponent of a particular global warming agenda, that being that global warming is a horrific imminent problem that will destroy the Earth very soon and that all steps need to be taken to stop it. What he is willing to do is to smear people and to misrepresent things to the media and to the public to get that message across. What's sad here is that there are partisan ties of his all the way up to the top of the Democratic Party, and he's using

those ties, and using his media connections, to push an agenda, a worst-case-scenario agenda of global warming, a sky-is-falling agenda of global warming, and anybody who is even perceived to disagree with him is labelled a censor and is demonized and vilified in the media, and the media of course is a willing accomplice here. . . . What you do have is hearsay coming from a handful of people who have clear partisan ties, and they're really coming after me as a Bush appointee and the rest of the Bush appointees because this is a partisan issue. It's a culture-war issue, they do not like Republicans, they do not like people who support the President, they do not like Christians, and if you're perceived to be disagreeing with them or being one of those people they will stop at nothing to discredit you.

It is yet another recurring theme that the Administration's technique, when confronted by criticism, is to accuse the critics of "just playing partisan politics", but it was a little startling to find it in this context – and as startling to discover that Deutsch regarded scientific conclusions as being somehow obedient to agendas. As the *New York Times* remarked in a February 9 editorial, "The shocker was not NASA's failure to vet Mr Deutsch's credentials, but that this young politico with no qualifications was able to impose his ideology on other agency employees."

It would be wrong to think, though, that the problem was merely one of a zealous young maverick. All this time, further horror stories were emerging from NASA about political manipulation of science, especially in the run-up to the 2004 presidential election. Bush had declared his great visionary goal was for the US to place a man on Mars, and the instructions were passed down through the NASA hierarchies that all NASA news releases should stress the contribution NASA scientists were making toward this "vision statement". Some of the results were ludicrous. In a December 2004 news release about research on wind patterns and warming in the Indian Ocean, JPL scientist Tong Lee was quoted as saying that some of the methods used in the study would be useful also in space exploration and studying the climate systems on other planets. Queried by his colleagues as to this statement, a startled Lee demanded that it be removed; NASA's press office duly removed it, but not NASA's public-affairs office in Washington, which retained it on the NASA website. And stories began emerging of how Gretchen Cook-Anderson, a NASA press officer whose job in 2004 was to manage the release of news in the earth sciences, had been pressurized by politically appointed senior members of NASA's hierarchy, notably Glenn Mahone, then Assistant Administrator for Public Affairs, to keep back until after the 2004 presidential election what might be seen as unfavourable news for the Administration's "the science is still unclear" pretence about global warming. Similarly, even through 2006 there was still a concerted political effort to expunge the phrase "global warming" from NASA sites and press releases in favour of the seemingly more neutral "climate change".

Finally, in late February 2006, NASA Administrator Michael D. Griffin began to conduct a review of NASA's communications policies that was hailed by scientists there as a genuine attempt to root out at least some of the political censorship and corruption of science that had become endemic in the agency. Hansen himself expressed cautious optimism. On March 14 Revkin reported him as saying: "The battle to achieve open communication between government scientists and their employer, the public, is far from won."

Seemingly so. In 2006 Pieter Tans and James Elkins, senior scientists at the Boulder, Colorado, laboratory of the US National Oceanic and Atmospheric Administration, recounted to federal investigators how between 2000 and February 2005, when Russia ratified the Kyoto Protocol, divisional director David Hofmann had a standing instruction that they were to make no mention of the word "Kyoto" during any of their presentations. Further, when in late 2005 Tans was organizing the Seventh International Carbon Dioxide Conference, he was instructed by Hofmann that the term "climate change" must not appear in the titles of any of the presentations. (Tans ignored the prohibition.) In response to the latter account, Hofmann said that he must have been misunderstood: he had merely been saying that a conference on global carbon-dioxide measurements should stick to exactly that, and not veer off into discussions of climate change!

Belatedly driven by the force of public opinion to admit that climate change was a reality and that something should be done about it, Bush pledged in late 2006 that increased research would be the centrepiece of his new climate strategy. Within weeks it emerged from a National Academy of Sciences study that over the previous two years NASA's earth-sciences budget had been reduced by 30%, with no plans for any reversal of the trend – indeed, the reckoning was that by 2010 the number of operative earth-observing instruments in US satellites would fall by 40%. This was in the context of a little-changed NASA budget: the funds that should have been spent on this critical work were instead being put toward preparation for a manned lunar base, a media-friendly scheme imposed on NASA by political fiat. While no one would object to the lunar-base project in principle, its status is far from urgent: moves to tackle climate change, by contrast, are. As a *New York Times* editorial summarized, "Studies that affect the livability of the planet seem vastly more consequential than completing a space station or returning to the Moon by an arbitrary date."

One crucial aspect in assessing global warming is assessing the earth's albedo (reflectivity); through knowledge of the albedo one can infer our planet's net energy balance. To this end, NASA built the Deep Space Climate Observatory, designed to be placed at the Lagrange-1 stationary orbit between the earth and the sun. At the time of writing, that instrument waits in a warehouse at Goddard, unlaunched and with no foreseeable prospects of being so.

The Administration's continued denial of the science concerning climate change was assisted in no small measure by the chairman of the Senate's Environment and Public Works Committee, Senator James Inhofe, who from 2003 until his replacement in early 2007 by Senator Barbara Boxer (b1940) – the Democrats having won both houses of Congress a few weeks earlier – used his position to promulgate the notion that global warming is simply a grandiose hoax put about by tree-huggers:* "The greatest climate

---

* This thesis was also advanced by novelist Michael Crichton (b1942) in *State of Fear* (2004). Alarmingly, in the wake of the novel's publication Crichton was called to the White House to consult with the President, as a supposed expert, on the "debate" concerning the science of climate change.

Senator James Inhofe (right)

threat we face may be coming from alarmist computer models." It is hard to work out if Inhofe was corrupting science for reasons of political expediency or simply through stupidity.

His claim continues to be that the earth is experiencing a natural warming trend that began around 1850, at the end of the four-century cool period known as the Little Ice Age, and that human activities have nothing to do with it; thus it's pointless for us to institute measures like capping greenhouse-gas emissions. To back up his case, Inhofe points to the period between the 1940s and the 1970s, when global temperatures fell slightly even as fossil-fuel consumption rose. This was certainly the case in the Northern Hemisphere, but represents only a minor fluctuation in a curve that has gone steadily upward since about the time of the Industrial Revolution; the earth's climate is an incredibly complex mechanism to which numerous factors contribute, so short-term fluctuations are to be expected in any of its trends. This is why there are disagreements in detail between the various computer models Inhofe decries as unreliable and alarmist: what he omits to mention is that they all agree to within a very narrow margin on the general picture – that human-generated global warming is a reality, and that the earth is rapidly approaching a climate crisis. Focusing on disagreements over the timing of the crisis is like arguing over whether one's speeding car is 10 or 20 metres from a cliff-edge.

For his swansong as chairman of the Senate's committee, Inhofe chose to hold on December 6 2006 a hearing concerning media coverage of climate change: laughably, he claimed the popular US news media were overhyping the subject, when a steady complaint from scientists and the informed public since before the dawn of the 21st century was that those media were largely ignoring it. This situation began to improve in 2006 largely due to a rise in public awareness brought about by the publicity surrounding the movie *An Inconvenient Truth* (2006), the climax of a campaign the movie's mainstay, Al Gore (b1948), had – despite much derision from rightist pundits – been waging for some 30 years: the mainstream news outlets could no longer get away with ignoring something that most of their audience were talking about.*

Inhofe's hearing had its hilarious moments. The geologist David Deming, a stalwart of the anti-environmentalist organization the National Center for Policy Analysis, began his closing statement thus:

> As far as I know, there isn't a single person anywhere on earth that's ever been killed by global warming. There is not a single species that's gone extinct. In fact, I'm not aware, really, of any deleterious effects whatsoever. It's all speculation.

The extinction over the past few years of numerous species owing to global warming is well documented – as a single example, over 70 species of tree frogs have gone. At the time of writing, there are major fears that polar bears may be driven to extinction within a matter of years as their Arctic habitat melts, and with it the bears' ability to find food. As for human deaths due to global warming, Deming presumably forgot about the tens of thousands of heat-related deaths in Europe, 10,000 in France alone, during the summer of 2003.

The political corruption of science under the Bush Administration has not been confined to the politicians themselves – far from it: propagandists in the broadcast media have played a major part in creating and maintaining the whole mess. Some are more egregious than others;

---

* Following the world premiere of the blockbuster movie *The Day After Tomorrow* (2004) in New York, Gore's slide show about global warming was seen by Laurie David, wife of TV comedian Larry David; Gore had been giving the show here and there all over the world since his disputed loss of the 2000 presidential election. Laurie David was sufficiently inspired by Gore's presentation and passion that she gathered a team to make a feature movie of it. In February 2007 *An Inconvenient Truth*, directed by Davis Guggenheim, received an Oscar.

listing the scientific distortions perpetrated by, say, radio
shoutmeister Rush Limbaugh (b1951) would fill a book.*
To cite just a single media incident as representative of a
myriad others, in the January 21 2006 issue of the TV series
The Journal Editorial Report, broadcast by Fox News and
linked to the *Wall Street Journal*, *WSJ* editorial page editor
Paul A. Gigot and his deputy, Daniel Henninger, discussed
a report recently released by the Max Planck Institute.
Henninger used it as an excuse to attack the Kyoto Accord:

> . . . the eminent Max Planck Institute in Heidelberg, Germany,
> has just reported in *Nature* magazine that plants, trees, forests
> emit 10% to 30% of the methane gas into the atmosphere.
> This is a greenhouse gas, the sort of stuff the Kyoto Treaty is
> meant to suppress. So, this is causing big problems for the
> tree-huggers, if plants, in fact, do cause greenhouse gases, and
> I have just one message for them: the next time you are out for
> a walk in the woods, breathe the methane.

Fair comment, one might suppose, were it not for the fact
that the report's *very first sentence* specifies that, because of
human activities, the atmospheric concentration of
methane has approximately tripled since the pre-industrial
era, in which context the 10–30% contribution of plants is
irrelevant; before the advent of industrialization, it was
obviously far higher in percentage terms (when human
contributions were virtually zero). The absolute level of
plant methane production has been gradually decreasing
during the same period, again thanks to human activities
(such as deforestation).

   Just to reinforce their point, the report's authors issued
a press release on January 18, three days before the
Gigot–Henninger distortion, in which they spelt out the
truth: "The most frequent misinterpretation we find in the
media is that emissions of methane from plants are respon-

---

* Such as *The Way Things Aren't: Rush Limbaugh's Reign of Error* (1995) by
Steve Rendall, Jim Naureckas and Jeff Cohen. Howlers include that it is
volcanoes, rather than manmade chlorofluorocarbons, that endanger
the ozone layer; that it has not been proven that nicotine is addictive;
that low levels of dioxin exposure aren't harmful; that the failure rate of
condoms in preventing HIV infection is as high as 20%; and that there
were fewer acres of forest land in the US in the time of the Founders
than there are now.

sible for global warming." They add that, while reforesta-
tion might trivially increase methane production, it at the
same time has the important beneficial effect of increasing
the amount of plant absorption of the far more dangerous
greenhouse gas carbon dioxide. While the excuse could be
made that Gigot and Henninger were guilty of no more
than bad journalism – failure to check facts – or plain ignorance, these seem unlikely given their positions at the *Wall
Street Journal*, where there are fact-checkers aplenty. The
most probable explanation is that their presentation was a
deliberate corruption of the scientific information for ideological purposes.

That same week, on January 25, CNN weatherman
Chad Myers put forward his own hypothesis as to what he
regards as the illusion of global warming. Temperatures
only *seem* to be rising as much as they are because buildings
are tending to spring up around the thermometers used by
climate scientists, and of course these suddenly mushroom-
ing metropolitan areas are warmer than the surrounding
countryside. Climate scientists are obviously too stupid to
take account of this effect. Myers did not expand on quite
how his hypothesis explains the increased melting of the
polar icecaps.

In despair at such stuff, in January 2007 Heidi Cullen, a
TV meteorologist and host of the programme *The Climate
Code*, called upon the American Meteorological Society to
withdraw its customary endorsement from those broadcast-
ing weather forecasters who tried to plant doubts in viewers'
minds about the reality of climate change. She pointed out
that if, for example, a meteorologist claimed on-air that
tsunamis were caused by the weather, the meteorologist
would be immediately recognized as incompetent and probably fired. Meteorologists who denied global warming
should be treated likewise.

Finally, in his State of the Union address of January 2007,
President Bush acknowledged for the first time the existence and threat of anthropogenic global warming. At the
time of writing he has yet to propose, or even support, any
realistic measures to do anything about it.

OUR GOD-GIVEN ENVIRONMENT

> Under Bush [as State Governor], Texas had the highest
> volume of air pollution, with the highest ozone levels of any
> state – while ranking forty-sixth in spending on environmen-
> tal problems. Moreover, after 1994 Texas was the nation's lead-
> ing source of greenhouse gases, accounting for 14 percent of
> the annual US total while boasting only 7 percent of the US
> population. Under Bush, Texas's oil refineries became the
> nation's dirtiest, with the highest level of pollution per barrel
> of oil processed. And because all such industrial effluvia are
> concentrated in or near the state's poorest neighborhoods,
> Bush's Texas also led the nation in the number of Title VI civil
> rights complaints against a state environmental agency – in
> this case, the Texas Natural Resource Conservation
> Commission (TNRCC), which Governor Bush staffed brazenly
> with staunch anti-environmentalists like Ralph Marquez, a
> veteran of Monsanto Chemical, and Barry McBee, of the pro-
> business law firm Thompson & Knight.
>
> Mark Crispin Miller, *The Bush Dyslexicon* (2001)

The name of the Bush Administration's Clear Skies
Initiative of 2003 deployed the technique of Newspeak in
order to disguise what it was in fact about: through replacing the Clean Air Act, introduced in 1963 and bolstered
several times, most recently in 1990 by the George H.W.
Bush Administration, the aim of the Clear Skies Initiative is
to loosen the controls on industrial polluters, thereby
inevitably increasing air pollution and the consequent sickness and death rates.

It's worth summarizing the achievements of the Clean
Air Act. Even before the Act's 1990 strengthening, the EPA
estimated the Act had saved 205,000 premature deaths
within the continental US. Further, "millions" of US citizens
had been spared illnesses ranging up to

> heart disease, chronic bronchitis, asthma attacks, and other
> serious respiratory problems. In addition, the lack of the
> Clean Air Act controls on the use of leaded gasoline would
> have resulted in major increases in child IQ loss and adult
> hypertension, heart disease and stroke.

In 1999 the EPA did a study of the effects of the 1990
bolstering of the Act and estimated that, in the single year
2010, 23,000 premature deaths would be prevented because

of it, not to mention 67,000 incidents of chronic and acute bronchitis and a whopping 1.7 million asthma attacks. For that one year, 4.1 million work days would be saved – an enormous contribution to the US economy.

The Bush Administration disliked the Clean Air Act for ideological reasons: the thrust of almost all of the Administration's environmental policies has been to reduce the legal obligations of business at the expense, if neccssary, of ordinary citizens. A particular concern of the Administration related to the regulations governing mercury emissions by coal-fired power plants; mercury is a neurotoxin. Accordingly, the White House watched carefully as the EPA compiled an advisory report, scheduled for release in 2002, which examined the effect of environmental factors on children's health. That report concluded 8% of US women aged 16–49 (i.e., of potential child-bearing age) had blood mercury levels that significantly increased the likelihood of any children they bore having deficient motor skills and reduced intelligence.

And this was before the introduction of the Clear Skies Initiative, which would further increase atmospheric mercury levels in the US! Accordingly, in May 2002, before the report's release, it was taken by the White House Office of Management and Budget (OMB) and the Office of Science and Technology Policy (OSTP) for "review". Nine months later, in February 2003, when it still hadn't emerged from that "review" process, it was leaked by someone at the EPA to the *Wall Street Journal* (a newspaper generally very much in the Bush Administration camp). It seems likely that the report – which was then hurriedly issued by an embarrassed Administration – would never have seen the light of day had it not been for the leak, and for the *Wall Street Journal*'s decision to run the story *in extenso*.

Clearly action had to be taken to ensure the EPA didn't commit such a *faux pas* again – putting scientific fact above political considerations. The chastened EPA accordingly produced in early 2004 a revised set of proposals on the regulation of power-plant mercury emissions. Almost immediately it was discovered that 12 paragraphs of the report had been copied, more or less verbatim, from a strategy document produced earlier by power-industry lawyers.

Horrified by the effect the Clear Skies Initiative was

likely to have on the nation's health, four senators – three Republicans and one Democrat – proposed a counter-measure that would tackle the problems of atmospheric carbon dioxide, nitrous and nitric oxides, sulphur dioxide and mercury. Their proposal was passed to the EPA for analysis in terms of costs and benefits. For months there was silence. Finally, in July 2003, another despairing EPA staffer leaked internal documents to the press, this time the *Washington Post*: the four senators' proposal would be more effective and speedier in reducing the pollutants than the Clear Skies Initiative, thereby saving some 18,000 lives by 2010, not to mention some $50 billion annually in health costs. The costs to industry of implementing the senators' proposal rather than the Clear Skies Initiative would be, the EPA concluded, "negligible". According to EPA staffers, this information was suppressed by Jeffrey Holmstead, Assistant Administrator for Air and Radiation.

Holmstead is yet a further example of the placing of appointees in positions within the science agencies for which they have no relevant qualifications except political loyalty; before his appointment, he was a lawyer, employed by the firm Latham & Watkins, which represented one of the US's largest plywood manufacturers. In January 2002 Holmstead, by now at the EPA, held a meeting of EPA staffers and the EPA Air Office's general counsel, William Wehrum, who by curious coincidence had previously been a partner in . . . Latham & Watkins. The purpose of the meeting was to discuss a rule governing emissions of formalde-hyde by wood-products plants. To represent the industry were Timothy Hunt, a lobbyist for the American Forest & Paper Association, and that organization's lawyer, Claudia O'Brien, who had earlier been a partner in . . . Latham & Watkins.

O'Brien recommended that supposedly low-risk prod-ucts should be exempt from any new emission controls, since the cost of introducing them would make US manu-facturers vulnerable to cheaper foreign competition. To the astonishment of the staffers, who knew that this proposal would violate the 1990 Clean Air Act provisions, Holmstead backed it.

So far as the Clear Skies Initiative and emissions by power plants were concerned, the Administration apparently decided EPA leakers might continue to bedevil White House adulteration of science, and thus introduced a new scheme. The New Source Review, by which power-plant emissions were gauged, would be "re-interpreted" – in other words, made less stringent. In 2002, Holmstead reassured a Senate Committee that the new, laxer rules would not be applied retroactively. Surprise, surprise, in November 2003 the top brass of the EPA announced to staffers that the rules would be applied retroactively: cases against about 50 coal-burning power plants guilty of violating the Clean Air Act were to be dropped. This was done at the instigation of the mysterious energy task force convened earlier by Vice President Richard M. Cheney (b1941); the precise person-nel of the task force has been stubbornly concealed from Congress and public alike, but it is known to have been staffed entirely by representatives of the power industry.

In February 2005 the EPA's Inspector General admitted that pressure had been put on the body's scientists to alter their reported results on the impact of mercury pollution to bring them into line with the industry-friendly conclusions the Administration demanded. That same month, the Government Accountability Office announced that indeed the EPA's results had been falsified so as to give the impres-sion that mercury poisoning's effects on brain development in fetuses and infants were significantly less than they in fact are. The EPA responded to these grave charges by making no changes at all to its final ruling, which was issued on March 15 2005.

Shortly afterwards, yet another piece of scientific fraud was discovered. A government-commissioned Harvard study had shown the costs of mercury pollution to be higher than previously thought, and the benefits of tighter control greater. The results of this study had been suppressed entirely by the EPA's political appointees.

In early 2006, a new measure to reduce environmental protection was a proposal to emasculate the Toxics Release Inventory program. This program, initiated during the aftermath of the 1986 disaster in Bhopal, India, requires companies to declare annually how they are disposing of some 650 different toxic chemicals in their wastes – in other words, where the poisons are going and how their environmental impact is being controlled. The new initiative would reduce the frequency of reporting from annually to biannually, would increase by a factor of 10 the threshold over which the reporting of released toxins had to be reported, and would remove altogether the obligation to report the release of cumulative toxins – such as lead and mercury – up to an annual level of about 230kg (500 pounds).

Unsurprisingly, the Attorneys General of 12 states were among the countless individuals and organizations who immediately protested: they have lives to save rather than profits to protect. After disasters like 2005's Hurricane Katrina, for example, it's vital that responders have ready access to information as to where toxic chemicals might have been released by the upheaval. Perhaps more surprisingly, companies and corporations were far from unanimous in welcoming the proposed relaxation of the rules, with many saying they would carry on reporting under the old rules anyway. Typical was the reaction of Edwin L. Mongan III, Director of Energy and Environment at DuPont: "It's just a good business practice to track your hazardous materials."

The American Medical Association no longer advises American delegations to UN summits on children's issues; Concerned Women for America does instead. Leaders of the National Association of People with AIDS no longer sit on the presidential AIDS advisory council, though religious abstinence advocates do; and members of the right-wing Federalist Society now vet judicial nominees rather than the mainstream American Bar Association.
Esther Kaplan, *With God on Their Side* (2004)

As noted, the ploy of making political appointments to supposedly scientific posts is endemic. Yet another example is James Connaughton, Chair of the Council on Environmental Quality, an advisory agency to the executive office of the President. His responsibility is "to bring together the nation's environmental, social, and economic priorities" and to "prepare the president's annual environmental quality report to Congress" – very important responsibilities indeed, especially in an era where the environment is, and hence human lives are, under greater threat than ever.

Connaughton is not an environmental scientist but a lawyer, and a lawyer who has countless times done battle with the EPA on behalf of big business. In 1993 he was the co-author of the article "Defending Charges of Environmental Crime – The Growth Industry of the 90s"; this appeared in *Champion Magazine*, published by the National Association of Criminal Defense Lawyers. He has lobbied on environmental issues on behalf of major corporations and corporate associations including the Aluminum Company of America, the Chemical Manufacturers Association, and General Electric. The latter is believed responsible for more toxic sites in the US than any other corporation.* ASARCO, another of Connaughton's clients, has lobbied that the US adhere to the 1942 drinking-water standard that permitted 50 parts per billion of arsenic. These are among many other corporate and trade association clients Connaughton has represented at the state, federal and international level, frequently acting as their defence lawyer in cases of environmental crimes.

It could of course be argued that Connaughton is qualified to be Chair of the CEQ on the grounds of his supreme knowledge of environmental law. It could equally be argued that Al Capone had a supreme knowledge of the criminal law, without necessarily regarding him as an ideal candidate for the office of Attorney General. Presumably, though, Connaughton passed the relevant political "litmus test" – an approval based on ideological loyalties rather than scientific qualifications. Bush-appointed Science Advisor John H. Marburger III (b1941), who *is* a scientist, has stated: "[T]he accusation of a litmus test that must be met before someone can serve on an advisory panel is preposterous." In fact, numerous eminent scientists have reported that exactly such a test has been applied to them when they were being considered for relevant posts within the Administration. The Union of Concerned Scientists has investigated these reports and detailed many of them in its various publications (see Bibliography) called *Scientific Integrity in*

* Although very recently it has begun to make welcome noises about cleaning up its act.

Policymaking. It seems Marburger must have been speaking on the basis of incorrect information.

Further curious appointees in the area of the environment have included Gale Norton (b1954) as Secretary of the Interior and Ann Veneman (b1949) as Agricultural Secretary. Both were protégées of the infamous James Watt (b1938), the anti-environmentalist appointed by Ronald Reagan as his Secretary of the Interior, and have worked for Watt's Mountain States Legal Foundation – self-described on occasion as "the litigation arm of Wise Use" (see page 233).

There was a solution to all these problems. In early 2007 Bush issued an executive order to ensure White House control, either directly or through political appointees, over the directives issued by the EPA and the Occupational Safety and Health Administration, among other federal agencies. It was one of his baldest moves yet to subject the findings of science to ideological manipulation. We can expect more.

NASTY! DIRTY! HORRID!

Sex education is yet another area bedevilled by the practice of installing political appointees. In his first term Bush appointed Claude Allen (b1960) as his deputy secretary of Health and Human Services; Allen's primary qualifications appeared to be that he opposed abortion and promoted sexual abstinence and Christian homeschooling. Kay Coles James was put in charge of the Office of Personnel; she had previously been a vice-president of the Family Research Council, a Christian right "traditional values" organization. Esther Kaplan, in *With God on Their Side* (2004), notes: "For a window into her worldview, one can look to her 1995 book *Transforming America from the Inside Out*, in which James likened gay people to alcoholics, adulterers and drug addicts." Another alumnus of the Family Research Council, Senator Thomas Coburn (b1948), was appointed joint chair of the presidential advisory council on HIV/AIDS. Robert Schlesinger, writing in *Salon* in September 2004, quotes Coburn as follows:

The gay community has infiltrated the very centers of power in every area across this country, and they wield extreme power . . . That agenda is the greatest threat to our freedom that we face today. Why do you think we see the rationalization for abortion and multiple sexual partners? That's a gay agenda.

Leaving aside the mystery of why gays should be so keen to promote abortion, Coburn, with his frequent anti-gay tirades, would seem an odd choice to offer objective advice on AIDS. (His views on abortion are odd, too: he has advocated the death penalty for doctors who've performed abortions despite himself having, as an obstetrician, performed abortions.)

But perhaps the most controversial of this wave of appointments was that of W. David Hager (b1946) in 2002 to the Advisory Committee for Reproductive Health Drugs of the Food and Drug Administration (FDA); according to *Time* magazine for October 5 2002, "Hager was chosen for the post by FDA senior associate commissioner Linda Arey Skladany, a former drug-industry lobbyist with longstanding ties to the Bush family." He is an obstetrician and gynaecologist with impressive conservative-Christian credentials – Focus on the Family, Christian Medical and Dental Society, etc. – is a tireless advocate of "traditional family values", and is author of books like *As Jesus Cared for Women: Restoring Women Then and Now* (1998).

Hager

Some of his medical views are idiosyncratic: in one of his books he recommends Bible study as a cure for menstrual cramps, and his patients have claimed (he denies this) that he refuses contraceptive prescriptions to unmarried patients. According to his own account, he was approached by the White House in 2001 and asked if he would be a possible candidate for the post of Surgeon

General. Later this was amended, and he was appointed to two advisory boards. Soon afterwards, he was asked to resign from those and instead join the FDA's reproductive drugs panel as its chairman; he quoted the officials as saying, "[T]here are some issues coming up we feel are very critical, and we want you to be on that advisory board."*

When his nomination as chairman of the FDA's reproductive drugs panel was announced in late 2002 there was an immediate outcry. In order to dodge the storm, the FDA used a frequent Administration trick: timing. It announced his appointment as a panel member (not chairman) on Christmas Eve, when not only was Congress out of session, and therefore unable to debate the issue, but the news was obscured from the public gaze by all the general media brouhaha over Christmas.

As for the opposition to his appointment, Hager has given this account:

> [T]here is a war going on in this country, and I'm not speaking about the war in Iraq. It's a war being waged against Christians, particularly evangelical Christians. It wasn't my scientific record that came under scrutiny, it was my faith. . . . By making myself available, God has used me to stand in the breach.

There followed the tortuous Plan B saga involving Hager. In December 2003 two scientific advisory committees at the Food and Drug Administration (FDA) voted unanimously that the emergency retroactive contraceptive pill Plan B, which had been approved as a prescription drug in 1999, was safe for over-the-counter sale, and by a majority of 23 to 4 that it should be made available for such sale. Its safety was attested to by about 70 scientific organizations, including the American Medical Association, the American Academy of Pediatrics, and the American College of Obstetricians and Gynecologists – not to mention the governmental scientific advisory bodies in the 33 countries in which it was already available over the counter.

---

* Just a few months earlier Hager had played an important role in the submission by the anti-abortion group Concerned Women for America of a "citizens' petition" demanded that marketing and distribution of the "abortion pill" RU-486 be halted.

Nevertheless, on May 6 2004 the Acting Director for the FDA's Center for Drug Evaluation and Research, Rear-Admiral Steven Galson, ruled that its open sale should continue to be prohibited. His stated grounds were that Plan B's manufacturers, Barr Pharmaceuticals, had failed to supply sufficient documentation showing the drug was suitable for use by teenagers under the age of 16 – this despite the fact that one of the two FDA scientific committees had extensively discussed exactly this point and decided there was no cause for concern; further, the committees had stressed that, in the case of very young girls, there are significant dangers involved in pregnancy, so its prevention was especially important. Of the veto Dr James Trussell, an FDA committee member and Director of Princeton University's Office of Population Research, said: "The objection . . . is nothing more than a made-up reason intended to sound plausible. From a scientific standpoint, it is complete and utter nonsense." He added: "Unfortunately, for the first time in history, the FDA is not acting as an independent agency but rather as a tool of the White House." Susan Wood, head of the Office of Women's Health, later resigned in protest, complaining that ideology had been allowed to trump science. John Jenkins, Director of the FDA's Office of New Drugs, commented: "The agency has not [previously] distinguished the safety and efficacy of Plan B and other forms of hormonal contraception among different ages of women of childbearing potential, and I am not aware of any compelling scientific reason for such a distinction in this case."

Jenkins was obviously searching the wrong area of human activity for his "compelling reason". To US right-wing Christian fundamentalists and the anti-abortion lobby, the idea of a morning-after pill was anathema, to the former because it would "obviously" unleash a torrent of promiscuity upon the land, to the latter because retroactive contraception is tantamount to abortion – the belief being that a fertilized ovum is a human being. (Despite some misleading news reports, Plan B is not in fact an abortifacient.)

According to Hager in a sermon he delivered in Wilmore, Kentucky, in October 2004, the person largely responsible for the scientific judgement being nullified was none other than himself. He claimed that, soon after the

23–4 vote, he was asked, as one of the four dissenters, to write a "minority report" outlining the reasons for rejecting the majority decision. In his own words:

> Now the opinion I wrote was not from an evangelical Christian perspective . . . I argued it from a scientific perspective, and God took that information, and he used it through this minority report to influence the decision. Once again, what Satan meant for evil, God turned into good.

An important question is: Who asked him for the minority report? The FDA does not commonly deal in such things, generally assuming that, if an overwhelming majority on a scientific advisory panel says something, it knows what it's talking about. Initially Hager told reporters and others that the request had come from within the FDA. Evidently realizing this could lead to a major political scandal, he soon backtracked, saying instead it had come from a mysterious "someone" outside the agency, and denying he'd ever stated otherwise. (Unfortunately for this latter claim, at least one journalist had kept Hager's e-mail.) The FDA likewise denied anyone within it had issued the request, claiming Hager's memo had been just a "private citizen letter".

Subsequent to Hager's 2004 boast, his ex-wife Linda, who had divorced him in 2002 after 32 years, decided to go public about why she had finally abandoned her marriage to a man who so publicly oozed family values – and with whom she had written the book *Stress and the Woman's Body* (1996). According to her, Hager had been sexually and emotionally abusive throughout their marriage; the details, which are quite horrific, were given in an article in *The Nation* for May 30 2005 by Ayelish McGarvey. As was forcefully pointed out by many, this was not a pedigree to recommend anyone for a position of power on women's health issues – and nor was Hager's frequently expressed view that women should regard men as the disciples did Jesus.

Finally, in August 2006, after nearly three years, the FDA caved in to public outrage and impartial science, permitting the over-the-counter sale of Plan B.

Once bitten, twice not shy, however, for the Administration. In November 2006 it appointed non-board-certified obstetrician/gynaecologist Eric Keroack (b1960) as Deputy Assistant Secretary for Population Affairs at the Department of Health and Human Services – i.e., as head of, among other things, the DHHS's family planning section, including the Title X program "designed to provide access to contraceptive supplies and information to all who want and need them, with priority given to low-income persons". His  previous appointments included serving as a medical adviser to the Abstinence Clearinghouse and as medical director of the Massachusetts pregnancy-counselling network A Woman's Concern: amid much pseudoscience,* this network opposes the provision of contraception *even to married women* – on the grounds that, inexplicably, contraception demeans women. Keroack also produced the pseudoscientific theory, unbacked by evidence, that women who over time have sex with a succession of different partners end up suffering an alteration in brain chemistry, through suppression of the hormone oxytocin, that makes it difficult for them thereafter to form long-term relationships. Oxytocin does indeed appear to have some effect on one's level of sociability, but the relationship is by no means a direct one: sometimes higher oxytocin levels make people grumpier, sometimes more amiable. There's no known relationship between oxytocin levels and marital/partnership happiness or number of sexual partners. After all, how could the oxytocin tell the marital status of a partner *vis-à-vis* the woman, or indeed one partner from another?

Ideologically opposed to contraception and a touter of pseudoscience and inaccurate medical information, Keroack was an odd pick as controller of federal funding in a scientific field whose focus is family planning – unless the

---

* For example, the myth that abortion increases the risk of breast cancer and the myth that sex education causes teen promiscuity. One of its pieces of advice to young people is that, on the (established) grounds that condoms reduce the risk of HIV transmission by 85%, "you have a 15% chance of contracting [HIV] while using a condom" without qualifications such as that this depends on whether or not your partner has HIV.

hidden aim was to run Title X and its affiliated programs into the ground. In April 2007 Keroack's past caught up with him and he abruptly resigned.

One means of distorting the nation's sex education emerged in 2002, when sharp-eyed observers discovered that the websites of the governmental health departments had been silently edited to accord more with the Administration's ideology. For example, where the site of the Centers for Disease Control and Prevention, concerned about the spread of AIDS and other sexually transmitted diseases, had previously stated that education on condom use did not increase youthful sexual activity and, far from encouraging an earlier onset of it, actually delayed it, that statement was found to have disappeared; furthermore, new information had been added emphasizing the risks of transferring STDs despite condom use:

> The surest way to avoid transmission of sexually transmitted diseases is to abstain from sexual intercourse, or to be in a long-term mutually monogamous relationship with a partner who has been tested and you know is uninfected. For persons whose sexual behaviors place them at risk for STDs, correct and consistent use of the male latex condom can reduce the risk of STD transmission. However, no protective method is 100 percent effective, and condom use cannot guarantee absolute protection against any STD.

There is no straightforward falsehood here, but the implication is distinctly different from that of a note found tucked away elsewhere on the site in a discussion of HIV: "The studies found that even with repeated sexual contact, 98–100 percent of those people who used latex condoms correctly and consistently did not become infected."

Likewise, the National Cancer Institute's site had once stressed there was no association between abortion and an increased risk of breast cancer in later life; the newly edited version claimed instead (falsely) that the science on the subject was "inconclusive". The results of an enormous Danish study, published in the *New England Journal of Medicine* in 1997, which found no relation at all between abortion and breast cancer, had been referred to approvingly on the earlier site; now there was no mention of it at all.

The adulteration of sex education is a major ideological issue; there's more on it in Chapter 5 (see pages 206ff).

YOU'RE JOKING, RIGHT?

There was considerable staff and public outrage when, of the 23 new books and other educational products considered on 2003 by officials at the Grand Canyon for sale at the National Park, the only one accepted, *Grand Canyon: A Different View* (2003) by Tom Vail, was a Creationist tract. This followed the suppression in 2001 (still continuing at the time of writing) by the leadership of the Park Service of a guidance leaflet for park rangers and other staff which decreed that any discussion of Creationism should stress its lack of scientific basis; in other words, stating that it is the responsibility of scientific institutions to teach science, not myth.

The superintendent of the Grand Canyon National Park, Joe Alston, attempted to have *Grand Canyon: A Different View* removed from sale at the park bookstores, but was promptly overruled by the head office of the National Park Service. However, the NPS's Chief of Communications, David Barna, announced there would be a high-level policy review of the issue; time showed that this was merely a publicity device to quell the growing protests, because at the end of 2006 the organization PEER (Public Employees for Environmental Responsibility) was able to discover, through a Freedom of Information request, that no such review had ever been initiated, let alone carried out . . . and the book remained on sale.

As one park geologist* put it, "This is the equivalent of Yellowstone National Park selling a book entitled *Geysers of Old Faithful: Nostrils of Satan*."

---

* Cited by Jeff Ruch, PEER's Executive Director. Ruch and PEER went further – too far, in fact, their December 2006 press release claiming that, for fear of upsetting Creationists, Grand Canyon rangers were under instruction to make no comment if asked about the canyon's age; this was a few weeks later exposed as false by Michael Shermer of the

(continued overleaf)

The falsehoods promulgated to the public, to Congress and even to the UN by the Bush Administration in the lead-up to the US invasion of Iraq in 2003 have been the subject of several books. Some of these false claims were of scientific or technological interest, perhaps most notoriously the case of the 100,000 high-strength aluminium tubes which the Hussein regime had apparently attempted to acquire on the international market: these, Bush told the UN in September 2002 and Colin Powell (b1937) repeated to the UN in February 2003, could only be to serve in gas centrifuges for enriching uranium – i.e., as a preliminary for the production of nuclear weapons. It seems the tight tolerances required for the tubes' dimensions and finish persuaded at least some in the CIA that they could have no other purpose.

Technical experts from the Oak Ridge, Livermore and Los Alamos laboratories of the US Department of Energy almost immediately dismissed the belief: the tubes were of the wrong dimensions for use in gas centrifuges. They were, however, identical to tubes Iraq had bought in the past for use in medium-range rockets. The State Department agreed. These scientific assessors also pointed out that, had the tubes been intended for the purpose stated by the Administration, there would be evidence also of the Hussein regime attempting to buy countless other essential and quite specific components for uranium enrichment.

Especially focused upon later, although media pundits and politicians alike shamefully ignored it at the time, was the reiteration of the aluminium-tubes falsehood by Bush himself in his 2003 State of the Union Address despite the fact that *just the previous day* the International Atomic Energy Agency (IAEA) had told the UN Security Council, which includes the US, that there was no evidence whatsoever of Iraq pursuing a nuclear-weapons program or of forbidden nuclear activities taking place at the country's relevant sites,

(continued) magazine *Skeptic*. Significant was how many individuals and organizations – including the cartoon strip *Doonesbury* – had in the interim been fooled by the PEER claim: it seemed so much of a piece with what else was going on that few thought to question it.

Intercepted aluminium tubes

and that the IAEA's analysis of the aluminium tubes had shown they would be useless for centrifuges.

There were other science-related dishonesties involved in the propaganda buildup to the invasion, most egregiously the continuation of the claim, long after it had been shown bogus, that Iraq had sought to buy yellowcake uranium from Niger. Also of interest were the insinuations about the as yet unsolved case of the anthrax parcels sent to various media and Democratic figures during the Fall of 2001. Could Hussein have been behind these attacks? Various of the Administration conceded this possibility, even though it was known to them that all the evidence pointed to a domestic perpetrator, probably a right-wing extremist in the Timothy McVeigh mould. The false insinuations about the anthrax packages jigsawed in nicely with the claims concerning Hussein's stupendous armoury of biological weapons, all of which ignored the fact that those weapons had been amassed prior to the first Gulf War, so most if not all would have degraded to near-uselessness by now.* The Administration's most Orwellian act of all, with the mainstream media as accomplice, was effectively to obliterate from US public awareness that the UN weapons inspectors led by Hans Blix (b1928), who were actually on the ground in Iraq, were clamouring that there were no signs of any active program there to create weapons of mass

* Many had in fact been sold to Hussein by the US for use in the earlier Iran-Iraq conflict.

destruction. After the attack Bush himself rewrote history on this issue, claiming that one motive for the invasion was that the Iraqi regime had barred the UN inspectors from the country.

The deeply unscientific War on Drugs, which was intended to reduce violent crime and which is still a vote-drawing catchphrase for any aspiring politician, was not originated by the Bush Administration, dating back to the Reagan Administration and perpetuated by successive Presidents since. The war's ineffectiveness is not surprising. A Columbia University survey done in 1997 of those convicted of violent crimes found that 3% had committed their crimes while high on cocaine and 1% were heroin addicts. These percentages might seem small but significant until we compare them with the equivalent figure for those who committed their violent crimes while drunk: 21%. Add in the equally lethal crime of drunk driving and the figure for alcohol-related mayhem would of course be even higher. Another threat targeted by the War on Drugs was the problem of "crack babies" – the offspring of crack-addicted mothers – who were widely reported in the late 1980s and early 1990s to suffer a syndrome comprising a psychological condition akin to autism, only worse, and a whole slew of physiological ailments; only later in the 1990s did attention reluctantly turn to the countless scientific reports that no such syndrome could be detected, and that the real danger to the developing fetus was the mother's consumption of drugs like tobacco and especially alcohol. Yet somehow the idea of a War on Alcohol rarely surfaces . . .

In the wake of the September 2001 attacks on the US, the War on Drugs lost most of its share of the limelight to the equally touted War on Terror. Again the scientific rationale was shaky: even in 2001, when some 3000 Americans lost their lives in the attacks, US citizens were more likely to die on the roads than as a result of terrorism. In any other year but 2001, the chance of being affected by terrorism in the US was considerably less than that of being struck by lightning. Yet fear, fanned by politicians and the media, gripped the population, and few chose to examine matters too carefully when, a couple of years later, the centrepiece of the War on Terror became the invasion of a country that had nothing to do with the 2001 attacks and, overall, however abominable its regime, less connection to international terrorism than the US itself, which during the Reagan Administration had financed terrorist organizations in South America and, it emerged in February 2007, under the Bush Administration was aiding Al Qaeda-linked terrorist organizations in Iran.

Other governments, in other times, have corrupted science for short-term political gain or for longer-term ideological reasons. In this long chapter we have looked at just three of the worst offenders. There are examples from earlier centuries of governments imposing their own scientific beliefs on the populace, but they're relatively rare outside theocracies: the political corruption of science is very much a modern phenomenon. There are, too, smaller-scale examples than these of modern governments attempting to pervert one or another aspect of science; one thinks immediately of the South African Government's mercifully brief flirtation with "traditional remedies" as a counter to the spread of AIDS. But such episodes are dwarfed by the systematic onslaught mounted by the three regimes treated here.

It might be expected that, with the internet offering ever-greater freedom of communication, such onslaughts would become progressively harder to mount. The experience of the Bush Administration has proven the opposite. Unless we, the public of any and every nation, maintain a constant vigilance, then we can expect authoritarian regimes everywhere to recognize the benefits – however illusory those benefits might in fact be – of corrupting science at its roots. If we let them get away with it, then we can indeed expect the arrival of, in every sense of the term, a new dark age.

BIBLIOGRAPHY

Alterman, Eric, and Green, Mark: *The Book on Bush: How George W. Bush (Mis)leads America*, Viking, 2004

Babbage, Charles: *Reflections on the Decline of Science in England, and on Some of Its Causes*, Fellowes, 1830

Bartholomew, Robert E., and Radford, Benjamin: *Hoaxes, Myths, and Manias: Why We Need Critical Thinking*, Prometheus, 2003

Berger, Peter L.: *The Sacred Canopy: Elements of a Sociological Theory of Religion*, Doubleday, 1967

Bernasconi, Robert (ed): *American Theories of Polygenesis*, Thoemmes, 2002

Berra, Tim: *Evolution and the Myth of Creationism: A Basic Guide to the Facts in the Evolution Debate*, Stanford University Press, 1990

Bloch, Sidney, and Reddaway, Peter: *Soviet Psychiatric Abuse: The Shadow Over World Psychiatry*, Gollancz, 1984

Block, N.J., and Dworkin, Gerald (eds): *The IQ Controversy: Critical Readings*, Pantheon, 1976

Brand, Stewart: *The Clock of the Long Now: Time and Responsibility*, Basic Books, 1999

Broad, William J.: *Teller's War: The Top-Secret Story Behind the Star Wars Deception*, Simon & Schuster, 1992

Broad, William, and Wade, Nicholas: *Betrayers of the Truth: Fraud and Deceit in the Halls of Science*, Simon & Schuster, 1982

Brock, David: *The Republican Noise Machine: Right-Wing Media and How It Corrupts Democracy*, Crown, 2004

Cirincione, Joseph: "A Brief History of Ballistic Missile Defense", Carnegie Endowment for International Peace, updated 2000 from 1998 conference paper "The Persistence of the Missile Defense Illusion"

Close, Frank: *Too Hot to Handle: The Race for Cold Fusion*, W.H. Allen, 1990

Cornwell, John: *Hitler's Scientists: Science, War and the Devil's Pact*, Viking Penguin, 2003

Dawkins, Richard: *Unweaving the Rainbow: Science, Delusion and the Appetite for Wonder*, Houghton Mifflin, 1998

Dennett, Daniel: *Darwin's Dangerous Idea: Evolution and the Meanings of Life*, Simon & Schuster, 1995

Dewdney, A.K.: *Yes, We Have No Neutrons: An Eye-Opening Tour through the Twists and Turns of Bad Science*, Wiley, 1997

Diamond, John: *Snake Oil, and Other Preoccupations*, Vintage, 2001

Diebold, John: *The Innovators: The Discoveries, Inventions, and Breakthroughs of Our Time*, Truman Talley/Dutton, 1990

Dowie, Mark: *Losing Ground: American Environmentalism at the Close of the Twentieth Century*, MIT Press, 1995

Dwyer, William M.: *What Everyone Knew About Sex*, Macdonald, 1973

Erickson, George A.: *Time Traveling with Science and the Saints*, Prometheus, 2003

Evans, Bergen: *The Spoor of Spooks*, Michael Joseph, 1955

Evans, Christopher: *Cults of Unreason*, Harrap, 1973

Eysenck, Hans: *Uses and Abuses of Psychology*, Penguin, 1953

Feder, Kenneth L.: *Frauds, Myths, and Mysteries: Science and Pseudoscience in Archaeology*, 3rd edition, Mayfield, 1999

Finkel, Elizabeth: *Stem Cells: Controversy at the Frontiers of Science*, ABC Books, 2005

Fitzgerald, A. Ernest: *The High Priests of Waste*, Norton, 1972

Flank, Lenny: *Deception by Design: The Intelligent Design Movement in America*, Red and Black, 2007

Forrest, Barbara, and Gross, Paul R.: *Creationism's Trojan Horse: The Wedge of Intelligent Design*, OUP, 2004

Fritz, Ben, Keefer, Bryan, and Nyhan, Brendan: *All the President's Spin: George W. Bush, the Media, and the Truth*, Touchstone, 2004

Futuyma, Douglas J.: *Science on Trial: The Case for Evolution*, Pantheon, 1982

Gardner, Martin: *Fads and Fallacies in the Name of Science*, Dover, 1952; revised and expanded edition, 1957

Gardner, Martin: *Science: Good, Bad and Bogus*, Prometheus, 1989

Gould, Stephen Jay: *The Mismeasure of Man*, Norton, 1981

Grant, John: *A Directory of Discarded Ideas*, Ashgrove Press, 1981

Grant, John: *Discarded Science*, FF&F, 2006

Gratzer, Walter: *The Undergrowth of Science: Delusion, Self-Deception and Human Frailty*, OUP, 2000

Haffner, Sebastian (trans Ewald Osers): *The Meaning of Hitler*, Harvard University Press, 1979

Heard, Alex: *Apocalypse Pretty Soon: Travels in End-Time America*, Norton, 1999

Hendricks, Stephenie: *Divine Destruction: Wise Use, Dominion Theology, and the Making of American Environmental Policy*, Melville House, 2005

Holton, Gerald: *Science and Anti-Science*, Harvard University Press, 1993

Huber, Peter W.: *Galileo's Revenge: Junk Science in the Courtroom*, revised edition, Basic Books, 1993

Ingram, Jay: *The Barmaid's Brain, and Other Strange Tales from*

Science, W.H. Freeman, 1998

Irvine, William: *Apes, Angels, and Victorians*, McGraw–Hill, 1955

Judson, Horace Freeland: *The Great Betrayal: Fraud in Science*, Harcourt, 2004

Kaminer, Wendy: *Sleeping with Extra-Terrestrials: The Rise of Irrationalism and Perils of Piety*, Pantheon, 1999

Kaplan, Esther: *With God on Their Side: How Christian Fundamentalists Trampled Science, Policy, and Democracy in George W. Bush's White House*, New Press, 2004

Kevles, Daniel J.: *The Physicists: The History of a Scientific Community in Modern America*, Knopf, 1978

Kitcher, Philip: *Abusing Science: The Case Against Creationism*, MIT Press, 1982

Koestler, Arthur: *The Case of the Midwife Toad*, Hutchinson, 1971

Kohn, Alexander: *False Prophets: Fraud and Error in Science and Medicine*, Blackwell, revised edition, 1988

Loewen, James W.: *Lies Across America: What Our Historic Sites Get Wrong*, Touchstone/Simon & Schuster, 1999

McRare, Ron: *Mind Wars: The True Story of Secret Government Research into the Military Potential of Psychic Weapons*, St Martin's Press, 1984

Martin, Brian: *Information Liberation: Challenging the Corruptions of Information Power*, Freedom Press, 1998

Mehta, Gita: *Karma Cola: Marketing the Mystic East*, Simon & Schuster, 1979

Millar, Ronald: *The Piltdown Men*, Gollancz, 1972

Miller, Arthur I.: *Empire of the Stars: Obsession, Friendship, and Betrayal in the Quest for Black Holes*, 2005

Miller, Mark Crispin: *The Bush Dyslexicon: Observations on a National Disorder*, Norton, 2001

Monbiot, George: *Heat: How to Stop the Planet Burning*, Allen Lane, 2006

Monmonier, Mark: *Drawing the Line: Tales of Maps and Cartocontroversy*, Henry Holt, 1995

Mooney, Chris: *The Republican War on Science*, Basic Books, 2005

Morgan, Chris, and Langford, David: *Facts and Fallacies: A Book of Definitive Mistakes and Misguided Predictions*, Webb & Bower, 1981

Morton, Eric: "Race and Racism in the Works of David Hume", *Journal on African Philosophy*, vol 1, no. 1, 2002

Numbers, Ronald L.: *The Creationists: The Evolution of Scientific Creationism*, Knopf, 1992

O'Donnell, Michael: *Medicine's Strangest Cases*, Robson, 2002

Park, Robert: *Voodoo Science: The Road from Foolishness to Fraud*,

Oxford University Press, 2000

Pennock, Robert T. (ed): *Intelligent Design Creationism and Its Critics: Philosophical, Theological, and Scientific Perspectives*, MIT Press, 2001

People for the American Way: *A Right Wing and a Prayer: The Religious Right and Your Public Schools*, People for the American Way, 1997

Porter, Roy: *Madness: A Brief History*, Oxford University Press, 2002

Porter, Roy, and Hall, Lesley: *The Facts of Life: The Creation of Sexual Knowledge in Britain 1650–1950*, Yale University Press, 1995

Ramos, Tarso: "Extremists and the Anti-Environmental Lobby: Activities Since Oklahoma City", Western States Center, 1997

Rampton, Sheldon, and Stauber, John: *Weapons of Mass Deception: The Uses of Propaganda in Bush's War on Iraq*, Tarcher/Penguin, 2003

Rees, Martin: *Just Six Numbers: The Deep Forces that Shape the Universe*, Weidenfeld & Nicolson, 1999

Regal, Brian: *Human Evolution: A Guide to the Debates*, ABC–CLIO, 2004

Rendall, Steve, Naureckas, Jim, and Cohen, Jeff: *The Way Things Aren't: Rush Limbaugh's Reign of Error*, Norton, 1995

Rice, Berkeley: *The C-5A Scandal: A $5 Billion Boondoggle by the Military–Industrial Complex*, Houghton Mifflin, 1971

Richelson, Jeffrey T.: *The Wizards of Langley: Inside the CIA's Directorate of Science and Technology*, Westview, 2001

Ronson, Jon: *The Men Who Stare at Goats*, Picador, 2004

Sagan, Carl: *Billions & Billions: Thoughts on Life and Death at the Brink of the Millennium*, Random, 1997

Sagan, Carl: *The Demon-Haunted World: Science as a Candle in the Dark*, Headline, 1996

Salomon, Jean-Jacques (trans Noël Lindsay): *Science and Politics*, MIT Press, 1973

Segal, Sanford L.: *Mathematicians Under the Nazis*, Princeton University Press, 2003

Scott, Eugenie C.: *Evolution vs. Creationism: An Introduction*, University of California Press, 2004

Shattuck, Roger: *Forbidden Knowledge: From Prometheus to Pornography*, St Martin's Press, 1996

Sheehan, Helena: *Marxism and the Philosophy of Science: A Critical History*, 2nd edition, Humanities Press International 1993

Shermer, Michael: *The Borderlands of Science: Where Science Meets Nonsense*, Oxford University Press, 2001

Shermer, Michael: *Why People Believe Weird Things: Pseudoscience, Superstition, and Other Confusions of Our Time*, W.H. Freeman, 1997

Shulman, Seth: *Undermining Science: Suppression and Distortion in the Bush Administration*, University of California Press, 2006

Sifakis, Carl: *American Eccentrics*, Facts on File, 1984

Taylor, John: *Science and the Supernatural: An Investigation of Paranormal Phenomena Including Psychic Healing, Clairvoyance, Telepathy and Precognition*, Dutton, 1980

Union of Concerned Scientists: "Scientific Integrity in Policymaking: An Investigation into the Bush Administration's Misuse of Science", Cambridge, Massachusetts, March 2004

Union of Concerned Scientists: "Scientific Integrity in Policymaking: Further Investigation of the Bush Administration's Misuse of Science", Cambridge, Massachusetts, July 2004

Vaidhyanathan, Siva: *The Anarchist in the Library: How the Clash Between Freedom and Control is Hacking the Real World and Crashing the System*, Basic Books, 2004

Vorzimmer, Peter J.: *Charles Darwin: The Years of Controversy* – The Origin of Species *and its Critics 1859–82*, University of London Press, 1972

Walker, Martin J.: *Dirty Medicine: Science, Big Business and the Assault on Natural Health Care*, Slingshot, 1993

Wanjek, Christopher: *Bad Medicine: Misconceptions and Misuses Revealed, from Distance Healing to Vitamin O*, Wiley, 2003

Weinberger, Sharon: *Imaginary Weapons: A Journey Through the Pentagon's Scientific Underworld*, Nation Books, 2006

Wertheim, Margaret: *Pythagoras' Trousers: God, Physics, and the Gender Wars*, Times Books, 1995

Wheen, Francis: *How Mumbo-Jumbo Conquered the World: A Short History of Modern Delusions*, Fourth Estate, 2004

Wynn, Charles M., and Wiggins, Arthur: *Quantum Leaps in the Wrong Direction: Where Real Science Ends . . . and Pseudoscience Begins*, Joseph Henry Press, 2001

Young, James Harvey: *The Medical Messiahs: A Social History of Health Quackery in Twentieth-Century America*, 2nd edition, Princeton University Press, 1992

INDEX